HOLT CALIFOR

Earth Science

Study Guide B
with Directed Reading Worksheets

HOLT, RINEHART AND WINSTON
A Harcourt Education Company
Orlando • Austin • New York • San Diego • London

TO THE STUDENT

Do you need to review the concepts in the text? If so, this booklet will help you. The *Study Guide* is an important tool to help you organize what you have learned from the chapter so that you can succeed in your studies. The booklet contains a Directed Reading worksheet and a Vocabulary and Section Summary worksheet for each section of the chapter.

Use these worksheets in the following ways:

- as a reading guide to identify and study the main concepts of each chapter before or after you read the text
- as a place to record and review the main concepts and definitions from the text
- as a reference to determine which topics you have learned well and which topics you may need to study further

Printed in the United States of America

ISBN-10: 0-03-099396-2
ISBN-13: 978-0-03099-396-1
3 4 5 6 7 018 11 10 09 08 07

Contents

The Nature of Earth Science

Tools of Earth Science

Earth's Systems and Cycles

Material Resources

Energy Resources

Plate Tectonics

Earthquakes

Volcanoes

Weathering and Soil Formation

Agents of Erosion and Deposition

Name ______________________________ Class ________________ Date ______________

Skills Worksheet

Directed Reading B

Section: Thinking Like a Scientist (pp. 8–15)

SCIENTIFIC HABITS OF MIND

______ **1.** What habits of mind do scientists share?
- **a.** Scientists are curious, skeptical, open minded, creative, and ethical.
- **b.** Scientists make mistakes but never let people know.
- **c.** Scientists try to find many ways that don't work.
- **d.** Scientists are curt, critical, open, and unethical.

______ **2.** What was biologist Jane Goodall curious about?
- **a.** where horses lived, what they ate, and how they interacted
- **b.** how fish are similar to humans
- **c.** where chimpanzees lived, what they ate, and how they interacted
- **d.** how birds lived for more than 30 years

______ **3.** What effect did Jane Goodall's questions, research, and writings have?
- **a.** It changed what scientists know about chimpanzees and other primates.
- **b.** It changed the way scientists think about humans and how they interact.
- **c.** It caused scientists to ask more questions about birds.
- **d.** It caused scientists to ask more questions about horses.

4. A habit of mind in which a person questions the truth of accepted ideas is called ______________________________.

5. What was biologist Rachel Carson skeptical of?

____

____

6. Carson's skepticism and questions encouraged people to do what?

____

____

7. What does keeping an open mind mean?

____

____

8. What does being creative help scientists do?

____

____

9. Upon what does science depend?

____

____

10. What happened to a group of scientists who claimed they had achieved cold fusion even though they had not?

____

____

____

11. To ensure honesty, what do scientists do with their work?

____

____

12. The process of having other scientists review scientific work before it is published is called ____________________.

13. What do scientists use to guide them as they conduct research?

____

____

14. When animals are involved in research, scientists must use ____________________.

15. When scientists involve people in research, what must the scientists do?

____

____

16. The process in which people are told of the risks of research and choose to participate is called ____________________.

17. When people involved in research are informed of the risks, what may they do?

____

____

____

WHAT DOES A SCIENTIST LOOK LIKE?

______ **18.** How is Stephen Hawking contributing to science?

a. He is a researcher who developed new treatments for the virus that causes AIDS.

b. He is adapting space technology to improve the lives of people in West Africa.

c. He is a theoretical physicist who has taught us about black holes in space.

d. He is studying where chimpanzees live, what they eat, and how they interact.

SCIENTIFIC LITERACY

19. The understanding of the methods of scientific inquiry, the scope of scientific knowledge, and the role of science in society is

called ______________________.

20. Name three skills that science can teach you.

____

____

____

21. Studying science will help you become a better-informed

______________________.

22. What kind of thinking skills do all successful scientists have?

____

23. What is the key to critical thinking?

____

____

SCIENCE IN OUR WORLD

______ **24.** Which of the following statements about science is NOT true?

a. Science is not something that happens only in the laboratory or in the classroom.

b. Science is a process and a way of thinking about the world.

c. Science happens only in the laboratory or in the classroom.

d. Ordinary people make important contributions to the advancement of science.

______ **25.** What happened in the 1990s as a result of the work of Mario Molina?
- **a.** The use of CFCs was banned in most of the world.
- **b.** Research to help reduce the effects of pollutants was banned.
- **c.** Chemicals were discovered that could repair the ozone layer.
- **d.** The use of CFCs was allowed in most of the world.

FROM THE CLASSROOM TO THE WORLD

26. What can students who take part in special science programs do?

____

____

27. What are three examples of activities that students can do at special science programs in California?

____

____

____

28. What do scientists working on the JASON Project do each year?

____

____

29. What two jobs might you be able to perform at your local science museum?

____

____

Name ______________________________ Class ______________ Date ____________

Skills Worksheet

Directed Reading B

Section: Scientific Methods in Earth Science (pp. 16–23)

LEARNING ABOUT THE NATURAL WORLD

1. What is the beginning of a process scientists use to learn more about the natural world?

____

WHAT ARE SCIENTIFIC METHODS?

2. A series of steps that scientists follow to answer questions and solve problems is called ______________________.

3. Scientists often change the number or order of ______________________ during an investigation.

ASKING A QUESTION

______ **4.** Which step helps scientists focus the purpose of an investigation?
- **a.** recycling resources
- **b.** asking a question
- **c.** using observation
- **d.** finding a dinosaur bone

5. What question might scientist David Gillette have asked when he examined some dinosaur bones?

____

____

6. David Gillette knew he would need to use ______________________ to answer his question about dinosaur bones.

FORMING A HYPOTHESIS

______ **7.** A possible explanation or answer to a question is a(n)
- **a.** prediction.
- **b.** data.
- **c.** variable.
- **d.** hypothesis.

______ **8.** To see if a hypothesis is true, someone must be able to
a. state it.
b. test it.
c. answer it.
d. guess it.

______ **9.** A statement in an if-then format is a(n)
a. prediction.
b. data.
c. variable.
d. question.

TESTING THE HYPOTHESIS

10. Pieces of information gathered through observation or experimentation are called ______________________.

11. A test of only one factor at a time is a(n) ______________________.

12. The factors in a controlled experiment are called ______________________.

13. How does changing one variable in a controlled experiment help a scientist?

____

____

14. Why must scientists keep accurate records of everything they do and observe?

____

____

15. Why must scientists use other methods besides controlled experiments to test a hypothesis?

____

____

____

16. If a large amount of data supports a hypothesis, what might scientists conclude about the hypothesis?

____

____

____

ANALYZING THE RESULTS

______ **17.** What is the next step for scientists after they finish their tests?

a. Make a prediction.
b. Analyze results.
c. Communicate results.
d. Draw a conclusion.

______ **18.** What might a scientist use to organize data?

a. a speech
b. a photograph
c. a graph
d. an experiment

DRAWING CONCLUSIONS

______ **19.** If a hypothesis is not supported,

a. the experiments were a waste.
b. the hypothesis was entirely correct.
c. scientists still may ask new questions and form new hypotheses.
d. scientists consider the experiment to be a success.

COMMUNICATING RESULTS

20. What are three ways that scientists can share what they have learned?

____

____

____

21. Why is sharing information important for scientists?

____

____

____

22. What are two factors that help maintain a scientist's believability?

____

____

23. What must scientists do when they learn that similar investigations had different results?

____

____

24. Even if there has been discussion about whether David Gillette's dinosaur is really from a new genus, his ______________________ continues today.

Name ______________________ Class ______________ Date ____________

Skills Worksheet

Directed Reading B

Section: Safety in Science (pp. 24–29)

THE IMPORTANCE OF SAFETY RULES

1. What are two purposes of safety rules?

____

____

2. What is the most important rule to keep you safe while doing science activities?

____

____

3. How can following safety rules after an accident help you?

____

____

____

ELEMENTS OF SAFETY

_____ **4.** What do safety symbols tell you?
- **a.** how to drive a car
- **b.** what to do to prevent injury or accidents
- **c.** how to do experiments
- **d.** when to leave the classroom

_____ **5.** What should you learn about each safety symbol?
- **a.** who made it up
- **b.** how to draw it
- **c.** where to find it
- **d.** what it warns you about

_____ **6.** What do lab procedures and instructions most closely resemble?
- **a.** football rules
- **b.** a treasure hunt
- **c.** a recipe
- **d.** a newspaper story

_____ **7.** If you can't complete some activity directions, you should
- **a.** keep on working and do what you think is correct.
- **b.** keep on working but ask your friend for help.
- **c.** stop working and start over.
- **d.** stop working and ask your teacher for help.

______ **8.** Why should you arrange your equipment and materials neatly during an experiment?
a. so you can easily locate the things you need
b. because it makes your work area look nice
c. because your teacher likes neatness
d. so you can finish more quickly

______ **9.** If you handle hot objects, you should
a. use your apron as a pot holder.
b. get someone else to hold them for you.
c. wear heat-resistant gloves.
d. stop working on the activity.

______ **10.** What should you do with extra or waste chemicals at the end of an activity?
a. Ask your lab partner what to do with them.
b. Follow your teacher's directions for disposal.
c. Wash them down the sink drain.
d. Take them home.

11. What are some rules for handling animals used in scientific research?

____

____

____

____

____

____

Match the correct example with the correct element of safety. Write the letter in the space provided.

______ **12.** wearing goggles and an apron

______ **13.** knowing what a picture of a hand holding a small animal means

______ **14.** washing glassware and checking it for chips

______ **15.** clearing books off the experiment work area

______ **16.** reading the instructions before starting a science activity

a. recognizing safety symbols
b. reading and following directions
c. practicing neatness
d. using proper safety equipment
e. proper clean-up procedures

PROPER ACCIDENT PROCEDURES

17. Why should you remain calm when an accident happens?

____

____

18. In the correct order, list the four steps to follow after an accident.

____

____

____

____

19. What are two things you should do if an accident happens?

____

____

20. What is first aid?

____

____

21. How can you help with first aid even if you are not trained to perform it?

____

____

____

Name ____________________ Class ____________ Date ____________

Skills Worksheet

Vocabulary and Section Summary B

Thinking Like a Scientist

VOCABULARY

After you finish reading the section, try this puzzle! Using each of these clues, fill in the blanks provided below with the letters of the word or phrase described below.

1. a person who uses observations and clear reasoning to understand processes and patterns in nature
2. a process in which people choose to participate in scientific research after they have been told of the risks involved
3. the practice of questioning the validity of accepted ideas or claims
4. the practice in which a scientist's work is looked at by other scientists before it is published
5. the understanding of the methods of scientific inquiry, the scope of scientific knowledge, and the role of science in society

1. ___ ___ ___ ___ ___ ___ ___ ___ ___
(blank 3 = 3, blank 5 = 21, blank 9 = 13)

2. ___ ___ ___ ___ ___ ___ ___ ___   ___ ___ ___ ___ ___ ___ ___
(blank 3 = 10, blank 7 = 22; second word: blank 1 = 2, blank 2 = 20, blank 6 = 5)

3. ___ ___ ___ ___ ___ ___ ___ ___ ___ ___
(blank 1 = 1, blank 3 = 7, blank 7 = 12)

4. ___ ___ ___ ___   ___ ___ ___ ___ ___ ___
(blank 2 = 4, blank 4 = 18; second word: blank 3 = 16, blank 5 = 11)

5. ___ ___ ___ ___ ___ ___ ___ ___ ___ ___
(blank 1 = 14, blank 4 = 17, blank 8 = 9, blank 10 = 6)

___ ___ ___ ___ ___ ___ ___ ___
(blank 4 = 15, blank 6 = 8, blank 8 = 19)

Discover the sentence below by filling in the blanks with the letter above the corresponding numbers.

6. ___ ___ ___ ___ ___ ___ ___   ___ ___ ___ ___ ___ ___ ___
1 2 3 4 5 6 7   8 9 10 11 12 13 14

___ ___ ___ ___ ___ ___ ___ ___.
15 16 17 18 19 20 21 22

SECTION SUMMARY

Read the following section summary.

- Scientists are curious, creative, skeptical, and open to new ideas.
- It is important for scientists to be honest and ethical in their treatment of humans and other living things.
- People from diverse backgrounds have made many contributions to the advancement of science.
- Increasing scientific literacy and developing critical-thinking skills are goals of science education.
- Scientists always evaluate the credibility of information that they receive.
- Scientists can have public roles in society. In addition to explaining scientific concepts to the media, scientists work to improve the quality of people's lives.
- There are many opportunities to participate in science programs in your community.

Name ______________________ Class ______________ Date ____________

Skills Worksheet

Vocabulary and Section Summary B

Scientific Methods in Earth Science

VOCABULARY

After you finish reading the section, try this puzzle! The underlined words are scrambled and are missing all their vowels. Write the completed words in the spaces provided.

1. A(n) TNLCRLD TRPXNM tests only one factor at a time by comparing a control group with an experimental group.

 ____

2. Any pieces of information obtained through observation or experimentation are known as TD.

 ____

3. To solve problems and answer questions, scientists follow a series of steps called FNSCTC SMDTH.

 ____

4. In a controlled experiment, only one factor, or BRVL, is tested at a time.

 ____

5. A testable idea or explanation that results in a scientific investigation is a(n) STYPHSH.

 ____

SECTION SUMMARY

Read the following section summary.

- Scientific methods are the ways in which scientists follow steps to answer questions and solve problems.
- The steps used in scientific methods are to ask a question, form a hypothesis, test the hypothesis, analyze the results, draw conclusions, and communicate results.
- A controlled experiment tests only one factor at a time so that scientists can determine the effects of changes to just that one factor.
- Accurate record keeping, openness, and replication of results are essential to maintaining a scientist's credibility.
- When similar investigations give different results, the scientific challenge is to verify by further study whether the differences are significant.

Name ______________________ Class ______________ Date ______________

Skills Worksheet

Vocabulary and Section Summary B

Safety in Science

VOCABULARY

After you finish reading the section, try this puzzle! Match each term with the correct definition. Write the letter in the space provided.

______ **1.** experiment

______ **2.** safety symbols

______ **3.** safety

______ **4.** first aid

______ **5.** hand safety

______ **6.** safety equipment

a. emergency medical care for someone who has been hurt or who is sick

b. signs that tell you what to do to prevent injury or accidents

c. freedom from injury, damage, or danger

d. the wearing of gloves to protect fingers from dangerous substances or sharp or hot objects

e. supplies or gear such as goggles, aprons, and gloves

f. a test or trial used to answer a question

SECTION SUMMARY

Read the following section summary.

- Following safety rules helps prevent accidents and helps prevent injury when accidents happen.
- Five elements of safety are recognizing safety symbols, following directions, being neat, using safety equipment, and using proper cleanup procedures.
- Animals used in scientific research require special care.
- When an accident happens, assess what happened, secure the area, report the accident, and help care for injuries or help clean up.
- First aid is emergency medical care for someone who has been hurt.

Name ______________________ Class ______________ Date ______________

Skills Worksheet

Directed Reading B

Section: Tools and Measurement (pp. 44–49)

1. Something that helps you do a task is called a(n) ______________________.

TOOLS FOR SCIENCE

2. Name a tool that you could use to measure the volume of water in a jar.

____

3. What do microscopes and magnifying lenses help you do?

____

____

4. What are the three main parts of a reflecting telescope?

____

____

____

Match the correct description with the correct term. Write the letter in the space provided.

______ **5.** used to measure mass

______ **6.** used to measure length

______ **7.** used to measure time

______ **8.** used to measure temperature

______ **9.** used to measure force

______ **10.** used to measure volume

a. meterstick

b. thermometer

c. graduated cylinder

d. spring scale

e. balance

f. stopwatch

11. Name three tools that can be used to analyze data.

____

____

____

MEASUREMENT

12. Give two examples of standardized units that were used for measurement hundreds of years ago.

____

13. Why are systems of measurement used in the past considered unreliable?

____

____

14. What are two names for the measurement system that was developed by the French Academy of Sciences between the late 1700s and the 1900s?

____

15. What are two advantages of using SI measurements?

____

____

16. One kilometer is equal to how many meters?

____

17. One gram is equal to how many kilograms?

____

18. When measuring temperature, 100°C equals how many kelvins?

____

19. The basic SI unit used to measure length is called a(n) ______________________.

20. A measure of the size of a surface or a region is called ______________________.

21. What is the equation used for calculating the area of a square or rectangle?

____

22. A measure of the amount of matter in an object is called ______________________.

23. The basic unit for mass is called a(n) ______________________.

24. Name two SI units that can be used to describe the mass of smaller objects.

____

____

25. Name one SI unit that can be used to describe the mass of large objects.

____

26. A measure of the size of a body or region in three-dimensional space is called ______________________.

27. How would you calculate the volume of a box-shaped object?

____

____

28. The volume of a liquid is often given in ______________________.

29. A measure of how hot or cold something is is called

______________________.

30. What is the SI base unit for temperature called?

____

WRITING NUMBERS IN SCIENTIFIC NOTATION

______ **31.** Shorthand that can be used to write very large numbers and very small numbers is called what?

a. Morse code
b. multiplication
c. scientific notation
d. calligraphy

Name ______________________ Class ______________ Date __________

Skills Worksheet

Directed Reading B

Section: Models in Science (pp. 50–55)

1. A pattern, plan, representation, or description designed to show the structure or workings of an object, system, or concept is called

   a(n) ______________________.

2. What are two tasks that models allow scientists to perform?

   ____

   ____

TYPES OF MODELS

3. What two types of things can models represent?

   ____

   ____

4. Models that you can touch are called ______________________.

5. What do physical models often look like?

   ____

6. What type of physical model most accurately represents Earth?

   ____

   ____

7. What type of physical model does not accurately represent Earth?

   ____

   ____

8. A model that is made up of mathematical equations and data

   is called a(n) ______________________.

9. Name two examples of something that could be calculated by using a simple mathematical model.

   ____

   ____

10. What tool is needed to process complex mathematical models?

   ____

11. How many calculations can supercomputers make per second?

____

12. Why are computers needed to make complex mathematical models?

____

13. Why do scientists who are trying to predict Earth's climate in 100 years use computers?

____

14. Computer models do not make exact predictions about future climates, but they estimate what might happen if ______________ change.

PATTERNS IN NATURE

______ **15.** It is possible for scientists to make models because
- **a.** models always follow patterns in nature.
- **b.** events in nature often follow predictable patterns.
- **c.** events in nature often follow unpredictable patterns.
- **d.** models are predictable but nature is unpredictable.

______ **16.** What is the basis of science?
- **a.** making models
- **b.** using computers
- **c.** observing patterns in nature
- **d.** designing experiments

THEORIES AND LAWS

______ **17.** What can observing patterns in the natural world lead to?
- **a.** publication of research without review
- **b.** development of scientific theories and laws
- **c.** unpredictable patterns in nature
- **d.** observations that are always accurate

18. A descriptive statement or equation that reliably predicts events under certain conditions is called a(n) ______________.

19. A system of ideas that explains many related observations and is supported by a large body of evidence acquired through scientific investigation is called a(n) ______________.

20. Why are scientific theories considered as useful and important as scientific laws?

____

____

21. What did scientists do when they observed planetary movements that did not fit the Earth-centered universe theory?

____

____

22. Sir Isaac Newton discovered the ________________ in 1665.

23. What does the law of universal gravitation state?

____

____

24. Scientists used the law of universal gravitation to strengthen the ________________ of a sun-centered solar system.

25. Why are all models limited?

____

____

26. Name two reasons why existing models can change.

____

____

27. What helps create new models that allow us to understand the world in a different way?

____

____

____

Name ______________________ Class ______________ Date ____________

Skills Worksheet

Directed Reading B

Section: Mapping Earth's Surface (pp. 56–63)

1. A flat representation of Earth's curved surface is called

a(n) ______________________.

FINDING DIRECTIONS ON EARTH

2. Earth's shape is best represented by a(n) ______________________.

3. Earth's axis of rotation can be used to establish ______________________.

USING A COMPASS

______ **4.** What does a compass use to show direction?
a. Earth's patterns of nature
b. Earth's natural magnetism
c. Earth's internal temperature
d. Earth's mass

______ **5.** The needle of a compass points toward the
a. magnetic north pole.
b. equator.
c. South Pole.
d. magnetic south pole.

FINDING LOCATIONS ON EARTH

______ **6.** Which of the following statements about latitude is NOT true?
a. The equator represents 0° latitude.
b. The North Pole is 90° north latitude.
c. The North Pole is 0° south latitude.
d. The South Pole is 90° south latitude.

Match the correct definition with the correct term. Write the letter in the space provided.

______ **7.** the line that represents 0° longitude

______ **8.** the imaginary circle halfway between the North and South Poles

______ **9.** the distance east and west from the prime meridian

______ **10.** the distance north or south from the equator

______ **11.** another name for a line of latitude

a. longitude
b. equator
c. latitude
d. prime meridian
e. parallel

12. What do lines of latitude and longitude look like on maps and globes?

____

INFORMATION SHOWN ON MAPS

______ **13.** The part of a map that shows symbols that need explanations is called a
- **a.** legend.
- **b.** scale.
- **c.** compass.
- **d.** title.

______ **14.** You can find information about the subject of a map in the
- **a.** date.
- **b.** title.
- **c.** projection.
- **d.** scale.

15. The relationship between the distance on Earth's surface and the distance on the map is the ______________________.

MODERN MAPMAKING

______ **16.** Data used in many maps are provided by
- **a.** projection.
- **b.** remote control.
- **c.** remote sensing.
- **d.** magnification.

______ **17.** The process of gathering information about an object without touching the object is called
- **a.** projection.
- **b.** remote sensing.
- **c.** mapping.
- **d.** modeling.

18. Which type of remote-sensing system records electromagnetic radiation using a satellite-mounted sensor?

____

19. Which type of remote-sensing system produces its own electromagnetic radiation and uses radar to gather data?

____

____

20. Orbiting satellites that send radio signals to a receiver on Earth make up the

______________________.

21. What three pieces of information does GPS calculate for a given location?

____

____

22. What do mapmakers use GPS for?

____

____

23. A computerized system that visually presents information about an area is called a(n) ______________________.

Name ______________________ Class ______________ Date ______________

Skills Worksheet

Directed Reading B

Section: Maps in Earth Science (pp. 64–69)

TOPOGRAPHIC MAPS

1. What type of map shows the surface features of Earth?

____

2. The height of an object above sea level is its ______________________.

CONTOUR LINES

Match the correct description with the correct term. Write the letter in the space provided.

_____ **3.** the difference in elevation between one contour line and the next

_____ **4.** a darker, heavier contour line; usually every fifth line

_____ **5.** a line that connects points of equal elevation

_____ **6.** the difference in elevation between the highest and lowest points of an area being mapped

a. contour line

b. contour interval

c. index contour

d. relief

7. Why are index contours used on topographic maps?

____

8. Contour lines that are close together show a(n) ______________________ slope.

9. Contour lines that are spaced far apart show a(n) ______________________ slope.

READING A TOPOGRAPHIC MAP

Match the correct description of a feature on a topographic map with the correct color. Write the letter in the space provided.

_____ **10.** wooded areas

_____ **11.** contour lines

_____ **12.** roads, bridges, and railroads

_____ **13.** major highways

_____ **14.** rivers, lakes, and oceans

a. black

b. green

c. brown

d. blue

e. red

THE RULES OF CONTOUR LINES

15. Contour lines never ______________________.

16. What does it mean when a contour line is shaped like a "V"?

____

____

____

17. What three things do closed circles represent on a topographical map?

____

____

____

18. How is a depression marked on a topographic map?

____

____

____

____

GEOLOGIC MAPS

Match the correct description with the correct term. Write the letter in the space provided.

______ **19.** map that records information such as rock units, structural features, mineral deposits, and fossil localities

______ **20.** rock of a given rock type and age range

______ **21.** location on a map where two geologic units meet

______ **22.** location of breaks in rock

______ **23.** topographic map upon which a geologist records bodies of rock and geologic structures

a. geologic map
b. fault
c. geologic unit
d. contact
e. base map

Name ______________________ Class ______________ Date ____________

Skills Worksheet

Vocabulary and Section Summary B

Tools and Measurement

VOCABULARY

After you finish reading the section, try this puzzle! In each statement below, the underlined word is incorrect. Write the correct term in the space provided.

1. Scientists often use a balance to measure area, which is the amount of matter in an object.

____

2. A measure of the size of a body or a region in three-dimensional space is called mass.

____

3. A measure of the size of a surface or a region is micrometer, which is based on two measurements.

____

4. The basic unit of length in the International System of Units is the inch.

____

5. A measure of the average kinetic energy of the particles in an object is volume.

____

SECTION SUMMARY

Read the following section summary.

- Scientists use tools to make observations, take measurements, and analyze data.
- Scientists must select the appropriate tools for their observations and experiments to take appropriate measurements.
- Scientists use the International System of Units (SI) so that they can share and compare their observations and results with other scientists.
- Scientists have determined standard ways to measure length, area, mass, volume, and temperature.
- Scientific notation is a way to express numbers that are very large or very small.

Name ______________________ Class ______________ Date ______________

Skills Worksheet

Vocabulary and Section Summary B

Models in Science

VOCABULARY

After you finish reading the section, try this puzzle! Use the clues to unscramble the letters in each word or phrase. Then, write the word or phrase in the space provided.

______________ **1.** explains many related observations and is supported by a large body of scientific evidence: eroyht

______________ **2.** reliably predicts events under certain conditions: wla

______________ **3.** allows you to calculate how far a car will travel: ahematlacimt dleom

______________ **4.** states that all objects in the universe attract each other with a force: alw fo vsaniurle vttnoiargai

______________ **5.** shows the structure or workings of an object, system, or concept: edoml

______________ **6.** often looks like the things it represents: hyspialc meldo

SECTION SUMMARY

Read the following section summary.

- Scientists must choose the right type of model to study a topic.
- Physical models, mathematical models, and computer models are common types of scientific models.
- Events in nature usually follow patterns. Scientists develop theories and laws by observing these patterns.
- Theories and laws are models that describe how the universe works. Theories and laws can change as new information becomes available.
- All models have limitations, and all models can change based on new data or new technology.

Name ______________________ Class ______________ Date ______________

Skills Worksheet

Vocabulary and Section Summary B

Mapping Earth's Surface

VOCABULARY

After you finish reading the section, try this puzzle! Use the clues below to write the term being described in the space provided. Then, find the term in the word search puzzle on the next page. Terms can be hidden in the puzzle vertically, horizontally, diagonally, or backward.

______________________ **1.** the distance east or west from the prime meridian

______________________ **2.** the process of gathering and analyzing information about an object without physically being in touch with it

______________________ **3.** a representation of the features of a physical body

______________________ **4.** the distance north or south from the equator

______________________ **5.** the line that represents 0° longitude

______________________ **6.** the imaginary circle that divides Earth into the Northern and Southern Hemispheres

D	M	I	R	K	W	D	M	C	N	U	S	P	E	A
P	R	I	M	E	M	E	R	I	D	I	A	N	R	Y
A	T	R	I	P	R	P	R	O	C	R	B	U	O	S
F	N	C	S	F	E	G	N	X	A	F	O	T	E	H
I	A	P	T	L	M	Z	M	I	Y	S	E	G	K	N
W	T	A	Q	O	O	E	D	U	T	I	T	A	L	U
C	M	Y	D	Z	T	N	N	L	F	Y	M	E	Q	K
S	N	X	E	K	E	H	G	V	P	H	G	V	M	V
G	S	M	H	G	S	B	Q	I	U	F	A	J	W	T
L	A	V	S	L	E	F	R	D	T	M	O	Q	U	O
P	W	B	J	M	N	R	I	E	T	U	V	M	C	P
C	D	W	V	N	S	P	N	S	X	E	D	D	D	M
Q	S	P	H	T	I	O	Y	H	L	J	G	E	W	R
I	O	C	A	E	N	T	C	Y	B	D	S	L	R	H
N	U	L	S	O	G	E	Q	U	A	T	O	R	S	E

SECTION SUMMARY

Read the following section summary.

- A map is a representation of the features of a physical body such as Earth.
- A compass is a tool that uses the natural magnetism of Earth to show direction.
- Latitude and longitude can be used to find points on Earth's surface.
- Most maps contain a title, a scale, a legend, an indicator of direction, and a date.
- Modern mapmakers use data gathered by remote-sensing technology to make most maps.
- Remote sensing is a way to collect information about an object without being in physical contact with the object.
- The global positioning system (GPS) calculates the latitude, longitude, and elevation of locations on Earth's surface.
- Geographic information systems (GISs) are computerized systems that allow mapmakers to store and use many types of data.

Name ______________________ Class ______________ Date ______________

Skills Worksheet

Vocabulary and Section Summary B

Maps in Earth Science

VOCABULARY

After you finish reading the section, try this puzzle! Use the clues below to solve the crossword puzzle.

ACROSS

3. the height of an object above sea level
4. a break in rock
5. rocks of a given rock type and age range
6. a map that shows the surface features of Earth

DOWN

1. a place where two geologic units meet
2. a representation that shows the distribution of geologic features in a given area
7. the variations in elevation of a land surface

SECTION SUMMARY

Read the following section summary.

- Contour lines connect points of equal elevation. They are used to show the shape of landforms.
- The contour interval is determined by the size and relief of an area.
- Geologic maps are designed to show the distribution of geologic features in a given area.
- Geologic units are the most important features shown on a geologic map.
- Geologic maps also show places where geologic units meet, where rocks are folded, and where rocks are broken.

Name ______________________ Class ______________ Date ____________

Skills Worksheet

Directed Reading B

Section: The Earth System (pp. 84–89)

EARTH: AN OVERVIEW

Match the correct description with the correct term. Write the letter in the space provided.

_____ **1.** the part of Earth that is water

_____ **2.** the part of Earth where life exists

_____ **3.** a mixture of gases that surround Earth

_____ **4.** the mostly solid, rocky part of Earth

a. geosphere
b. biosphere
c. hydrosphere
d. atmosphere

5. What four spheres do energy and matter flow through and between?

____

____

____

____

GEOSPHERE

Match the correct description with the correct term. Write the letter in the space provided.

_____ **6.** the central part of Earth; made of iron and nickel

_____ **7.** the thin, outermost layer of Earth

_____ **8.** the layer of rock between Earth's crust and core

a. crust
b. mantle
c. core

9. The physical layer of Earth that is divided into tectonic plates is called the

____________________.

10. The solid, plastic layer of Earth upon which tectonic plates move is called

the ____________________.

11. The lower, solid layer of the mantle is called the ____________________.

12. What is the outer core of Earth made of?

____

13. What is the inner core of Earth made of?

____

THE ATMOSPHERE

_____ **14.** How would the atmosphere best be described?
a. the inner layer of Earth's mantle
b. the mixture of invisible gases that surround Earth
c. the mixture of visible gases near Earth's surface
d. the rigid, outer layer of Earth's crust

_____ **15.** Where are most of Earth's atmospheric gases found?
a. about 500 km from Earth's surface
b. between 50 and 100 km from Earth's surface
c. within 8 to 12 km of Earth's surface
d. less than 8 km above Earth's surface

_____ **16.** In which layer of the atmosphere do we live?
a. mesosphere
b. stratosphere
c. thermosphere
d. troposphere

_____ **17.** The layer directly above the troposphere is called the
a. mesosphere.
b. stratosphere.
c. thermosphere.
d. hydrosphere.

_____ **18.** The coldest layer of the atmosphere is called the
a. mesosphere.
b. thermosphere.
c. stratosphere.
d. troposphere.

_____ **19.** The uppermost layer of the atmosphere is called the
a. mesosphere.
b. stratosphere.
c. thermosphere.
d. troposphere.

_____ **20.** Starting with the layer closest to Earth's surface, what is the order of layers in the atmosphere?
a. mesosphere, stratosphere, thermosphere, troposphere
b. stratosphere, mesosphere, thermosphere, troposphere
c. troposphere, stratosphere, mesosphere, thermosphere
d. thermosphere, mesosphere, stratosphere, troposphere

______ **21.** Which of the following is the main source of energy that reaches Earth?

a. volcano
b. lightning
c. sun
d. moon

______ **22.** Why does air in the atmosphere move?

a. because solar radiation heats Earth's surface unevenly
b. because solar radiation heats Earth's water unevenly
c. because solar radiation heats Earth's surface evenly
d. because solar radiation heats Earth's water evenly

______ **23.** Cold air sinks and forces warm air out of the way because

a. warm air contains more oxygen than cold air does.
b. warm air contains less oxygen than cold air does.
c. cold air is less dense than warm air is.
d. cold air is denser than warm air is.

______ **24.** Which of the following distributes energy throughout the atmosphere?

a. the movement of air
b. the movement of clouds
c. the movement of ocean waves
d. the movement of tectonic plates

______ **25.** The transfer of energy, especially heat, due to the movement of matter is called

a. deposition.
b. convection.
c. compression.
d. erosion.

THE HYDROSPHERE

26. What is the hydrosphere made of?

____

____

27. How much surface area is covered by the global ocean?

____

28. How much of Earth's water does the global ocean hold?

____

29. Why does the temperature of ocean water vary?

____

30. What two factors affect the density of ocean water?

____

____

31. Any movement of matter that results from differences in density is called a(n) ______________________

32. What is energy in the ocean distributed by?

____

THE BIOSPHERE

33. What is Earth's biosphere made up of?

____

34. Name the three parts of Earth that make up the biosphere.

____

____

____

35. What two factors are very important for the survival of living things?

____

____

36. What kind of environment must most plants and animals live in?

____

37. Plants and algae get their energy from the ______________________.

38. Energy enters the biosphere as ______________________.

39. Name two ways that energy is passed between organisms.

____

____

Match the correct description with the correct term. Write the letter in the space provided.

______ **40.** an organism that breaks down the remains of dead organisms

______ **41.** the process by which plants turn energy from the sun into chemical energy

______ **42.** a material used by plants to make food

a. carbon dioxide

b. photosynthesis

c. decomposer

Name ______________________ Class ______________ Date ______________

Skills Worksheet

Directed Reading B

Section: Heat and Energy (pp. 90–97)

WHAT IS TEMPERATURE?

1. Define *temperature.*

____

2. When particles are in motion, they have ______________.

3. The faster particles move, the more ______________ they have.

4. What does the temperature of a substance depend on?

____

5. The more kinetic energy that the particles of an object have, the higher the object's ______________.

6. Why do particles have different amounts of kinetic energy?

____

____

7. When you measure an object's temperature, you measure the ______________ of all the particles in the object.

8. Why would the temperature of a full teapot of tea and the temperature of a full teacup of tea be the same, even though the teapot holds more?

____

____

____

THERMAL EXPANSION

______ **9.** What happens to particles in a substance when the substance's temperature increases?

a. The particles all move at the same speed.
b. The particles move faster and move apart.
c. The particles move slower and move together.
d. The particles stop moving.

______ **10.** What happens to a substance when the space between the substance's particles increases?
a. The substance's kinetic energy decreases.
b. The substance's temperature decreases.
c. The substance evaporates.
d. The substance expands.

______ **11.** The increase in volume that results from an increase in temperature is called
a. thermal expansion.
b. compression.
c. convection.
d. density.

12. A hot-air balloon rises and a thermometer measures temperature through what process?

____

WHAT IS HEAT?

______ **13.** What is heat?
a. the movement of matter due to differences in density
b. a measure of a substance's total kinetic energy
c. the energy transferred between objects that are at different temperatures
d. a measure of the number of particles in a substance

14. Energy is always passed from an object with a(n) ______________ temperature to an object with a(n) ______________ temperature.

15. Define *thermal energy.*

____

16. What is the unit used to measure thermal energy?

____

17. An object with a high temperature has more ______________ than an object with a lower temperature does.

18. The more particles in a substance at a given temperature, the greater the substance's ______________.

19. What happens to the temperatures of a warm object and a cool object when the objects come into contact with each other?

____

____

20. When two objects with the same temperature touch each other, what happens to the thermal energy of both objects?

____

____

HOW IS HEAT TRANSFERRED?

Match the correct definition with the correct term. Write the letter in the space provided.

_____ **21.** the transfer of energy as heat through a material

_____ **22.** the movement of matter due to differences in density; the transfer of energy due to the movement of matter

_____ **23.** the transfer of heat or other energy as electromagnetic waves, such as visible light or infrared waves

a. radiation

b. convection

c. conduction

24. How does radiation differ from conduction and convection?

____

____

STATES OF MATTER

25. What are the three physical forms in which a substance can exist?

____

____

____

26. Name the four factors upon which a substance's state depends.

____

____

____

Match the correct description with the correct term. Write the letter in the space provided.

______ **27.** occurs when a substance changes from gas to liquid

______ **28.** occurs when a substance changes from liquid to solid

______ **29.** occurs when a substance changes from liquid to gas

a. freezing

b. evaporating

c. condensing

30. A change of state involves a transfer of ______________________ from one substance to another.

31. When a substance changes state, ______________________ is released or added to the substance.

32. A substance gains energy during what two processes?

____

33. A substance loses energy during what two processes?

____

Name ______________________ Class ______________ Date ______________

Skills Worksheet

Directed Reading B

Section: The Cycling of Energy (pp. 98–103)

THE FLOW OF ENERGY

1. What are three ways that energy can be carried from one place to another?

____

____

____

2. The transfer of energy from a warmer object to a cooler object is called

______________________.

3. Water waves, light waves, and sound waves transfer

______________________ through vibrations.

4. What are the two sources of energy for the Earth system?

____

____

5. Name the three processes that move energy through the geosphere, hydrosphere, biosphere, and atmosphere.

____

____

____

RADIATION

______ **6.** How much of the energy that reaches Earth is supplied by the sun through radiation?

a. 50%
b. 25%
c. 1%
d. 99%

Match the correct description with the correct term. Write the letter in the space provided.

______ **7.** energy that travels from the sun in waves

______ **8.** the full range of wavelengths, including radio waves, gamma rays, and visible light

a. electromagnetic radiation

b. electromagnetic spectrum

9. After energy from the sun is absorbed by Earth's atmosphere, geosphere, and hydrosphere, the energy is changed into ______________________.

10. Thermal energy is transferred through Earth's systems by which two processes?

____

____

CONVECTION

11. What is the role of convection in Earth's systems?

____

____

12. The movement of matter that results from differences in density is called a(n) ______________________.

Match the correct description with the correct term. Write the letter in the space provided.

______ **13.** where convection currents form when cold air sinks and forces warm air away from Earth's surface

______ **14.** where convection currents cause the movement of tectonic plates

a. mantle

b. atmosphere

CONDUCTION

15. Why does a warmer substance have more kinetic energy than a cooler substance?

____

____

16. When particles in a warm substance transfer energy to particles in a cooler substance, how is the temperature of the cooler substance affected?

____

____

17. Between which two spheres can energy be transferred by conduction?

____

____

18. At what point is energy passed to the atmosphere by conduction?

____

____

19. What condition must exist in order for energy to flow from the atmosphere to Earth?

____

____

EARTH'S ENERGY BUDGET

______ **20.** The four spheres of Earth are called *open systems* because energy

- **a.** can never be transferred between the spheres.
- **b.** is constantly exchanged between the spheres.
- **c.** is created in the atmosphere and destroyed in the geosphere.
- **d.** cannot be converted into other forms of energy.

______ **21.** The movement of energy between spheres can be thought of as an "energy budget" because Earth's energy

- **a.** will eventually be used up.
- **b.** can be added in one sphere and subtracted in another.
- **c.** must be equally balanced in the atmosphere and the geosphere.
- **d.** can be created and destroyed.

Name ______________________ Class ______________ Date ______________

Skills Worksheet

Directed Reading B

Section: The Cycling of Matter (pp. 104–111)

1. Why must matter cycle continuously on Earth?

____

THE CHANGING EARTH

2. What happens to matter as it cycles through the Earth system?

____

3. Give one example of a change on Earth that takes millions of years.

____

4. What is one thing on Earth that can change from one hour to the next?

____

THE ROCK CYCLE

Match the correct description with the correct term. Write the letter in the space provided.

______ **5.** the process by which rock changes from one form to another

______ **6.** the process by which rock is broken down by wind, water, and temperature changes

______ **7.** the process by which water, wind, ice, and gravity move rock from one place to another

a. weathering

b. erosion

c. rock cycle

8. An igneous rock may become a sedimentary rock as a result of which two processes?

____

____

9. What factor determines the forces that will act on a rock?

____

10. Where on Earth is rock exposed to weathering and erosion?

____

11. What are two forces that rock is exposed to deep inside Earth?

____

____

CLASSES OF ROCKS

12. Small pieces of rock that become cemented together form what type of rock?

____

13. After hot, liquid rock called magma cools and solidifies, it becomes

_______________________.

14. What type of rock forms as a result of chemical processes or changes in temperature and pressure?

____

15. Rock or mineral fragments that are not yet cemented together are called

_______________________.

16. How do clastic sedimentary rocks form?

____

____

17. How do chemical sedimentary rocks form?

____

____

18. How do organic sedimentary rocks form?

____

____

19. What two characteristics can be used to divide igneous rocks into groups?

____

____

20. What characteristic determines the chemical composition of an igneous rock?

____

21. Intense heat, pressure, or chemical processes result in the formation of

_______________________ rock.

22. Where do most metamorphic changes happen?

____

23. What is the difference between foliated metamorphic rock and nonfoliated metamorphic rock?

____

____

THE WATER CYCLE

24. Define *water cycle.*

____

____

25. What is the major source of energy that powers the water cycle?

____

Match the correct description with the correct term. Write the letter in the space provided.

______ **26.** the process by which liquid water changes into gaseous water vapor

______ **27.** the process by which water vapor is released from plants into the air

______ **28.** the process by which a gas, such as water vapor, turns into liquid water droplets

______ **29.** the process by which water droplets fall to Earth as rain

______ **30.** water that moves over the land surface

______ **31.** water that moves downward by gravity through spaces in soil and rock

a. precipitation
b. runoff
c. condensation
d. transpiration
e. groundwater
f. evaporation

THE CARBON CYCLE

Match the correct description with the correct term. Write the letter in the space provided.

______ **32.** the movement of carbon from the nonliving environment into living things and back

______ **33.** the process by which organisms break down the remains of dead organisms

______ **34.** the process of burning fossil fuels

a. combustion
b. carbon cycle
c. decomposition

THE NITROGEN CYCLE

______ **35.** What role do certain bacteria play in the nitrogen cycle?
a. Bacteria change nitrogen into groundwater.
b. Bacteria change nitrogen into a form that plants can use.
c. Bacteria change nitrogen into gaseous water vapor.
d. Bacteria change nitrogen into liquid water.

THE PHOSPHORUS CYCLE

______ **36.** Which of the following is NOT a way that phosphorus cycles through Earth?
a. Plants return phosphorus to the soil through transpiration.
b. Plants absorb phosphorus from the soil.
c. Phosphorus returns to the soil through decomposition.
d. Animals obtain phosphorus by eating plants.

OTHER CYCLES IN NATURE

______ **37.** When a living thing dies,
a. substances in its body cannot be recycled.
b. only the nitrogen in its body is recycled.
c. every substance in its body is recycled.
d. only the phosphorus in its body is recycled.

Name ______________________ Class ______________ Date ____________

Skills Worksheet

Vocabulary and Section Summary B

The Earth System

VOCABULARY

After you finish reading the section, try this puzzle! Use the clues below solve the following crossword puzzle.

ACROSS

3. any movement of matter that results from differences in density; may be vertical, circular, or cyclical

4. the central part of Earth below the mantle

DOWN

1. the layer of rock between Earth's crust and core

2. the movement of matter due to differences in density and the transfer of energy that results from this movement

3. the thin and solid outermost layer of Earth above the mantle

SECTION SUMMARY

Read the following section summary.

- The four divisions of Earth are the hydrosphere, atmosphere, geosphere, and biosphere.
- The geosphere is divided into layers based on composition and physical properties.
- Convection moves energy through the atmosphere and through the hydrosphere.
- Energy in the biosphere is transferred from the sun to plants and then from one organism to another.

Name ______________________ Class ______________ Date __________

Skills Worksheet

Vocabulary and Section Summary B

Heat and Energy

VOCABULARY

After you finish reading the section, try this puzzle! Use the clues below to write the term described in the space provided. Then, find the words in the word search puzzle on the following page. Words are hidden vertically, horizontally, diagonally, and backward.

______________________ **1.** a measure of how hot or cold something is; specifically, a measure of the average kinetic energy of the articles in an object

______________________ **2.** the energy transferred between objects that are at different temperatures

______________________ **3.** the kinetic energy of a substance's atoms

______________________ **4.** the transfer of energy as heat through a material

______________________ **5.** the movement of matter due to differences in density and the transfer of energy that results from this movement

______________________ **6.** the transfer of energy as electromagnetic waves

P	C	J	N	O	E	M	M	M	W	G	M	V	B	E	H	T	I
K	D	A	G	C	H	R	D	G	Y	F	N	C	K	T	M	U	F
M	F	Y	O	T	A	Z	U	H	P	O	N	T	B	A	Q	R	T
D	X	E	Q	Y	P	W	L	T	I	F	D	K	N	A	U	I	K
J	Z	D	H	X	B	B	I	T	A	Q	Q	D	Z	L	F	S	P
C	B	S	G	W	B	G	C	L	U	R	U	R	O	U	D	U	A
F	A	H	B	B	O	E	B	B	O	Q	E	D	E	O	Z	N	F
U	L	J	A	X	V	G	W	Z	Z	J	V	P	O	F	A	O	B
M	F	M	P	N	Y	M	A	G	J	L	Q	Y	M	V	N	I	Q
B	Z	A	O	Z	R	A	D	I	A	T	I	O	N	E	L	T	B
S	Z	C	T	B	Y	U	Y	R	R	G	W	J	O	R	T	C	P
Z	I	L	H	H	X	N	A	K	Z	M	X	R	K	P	R	U	I
S	P	F	V	I	Q	N	W	T	F	L	M	X	X	H	Q	D	Q
C	D	O	X	H	R	Q	S	H	S	W	L	B	T	V	N	N	B
T	X	M	M	J	K	X	Q	Y	X	Z	H	X	Y	T	H	O	U
E	R	Y	L	Q	W	X	C	M	F	O	J	E	X	K	R	C	B
Y	G	R	E	N	E	L	A	M	R	E	H	T	A	C	H	G	N
S	K	F	L	L	I	H	Z	V	P	U	B	K	R	T	H	M	I

SECTION SUMMARY

Read the following section summary.

- Heat moves from warmer objects to cooler objects until all of the objects are at the same temperature.
- Conduction is the transfer of energy as heat through a solid material.
- Convection is the transfer of energy due to the movement of matter.
- Radiation is the transfer of energy as electromagnetic waves. Radiation differs from conduction and convection because radiation can transfer energy through empty space.
- A substance's state of matter depends on the speed of the particles in the substance. Changes of state result from the transfer of energy.

Name ______________________ Class ______________ Date ____________

Skills Worksheet

Vocabulary and Section Summary B

The Cycling of Energy

VOCABULARY

After you finish reading the section, try this puzzle! Unscramble the letters at the end of each description to find the word that is being described.

1. another term for heat transfer; the transfer of energy from a warmer object to a cooler object: TAHE OFLW

___ ___ ___ ___ ___ ___ ___ ___

2. all the frequencies or wavelengths of electromagnetic radiation: MECELORTGIENCTA RUPMTECS

___ ___ ___ ___ ___ ___ ___ ___ ___ ___ ___ ___ ___ ___ ___

___ ___ ___ ___ ___ ___ ___ ___

3. the transfer of energy as electromagnetic waves: OIANDIATR

___ ___ ___ ___ ___ ___ ___ ___ ___

4. any movement of matter that results from differences in density; may be vertical, circular, or cyclical: ONCOECTNIV RRNTUEC

___ ___ ___ ___ ___ ___ ___ ___ ___ ___

___ ___ ___ ___ ___ ___ ___

5. the transfer of energy as heat through a material: DNTCOIUONC

___ ___ ___ ___ ___ ___ ___ ___ ___ ___

SECTION SUMMARY

Read the following section summary.

- Energy can be transferred from one place to another by heat flow, by waves, or by objects that are moving.
- Heat flow is the transfer of energy from a warmer object to a cooler object.
- Energy from the sun reaches Earth by radiation.
- Energy is transferred through the oceans, the atmosphere, and the geosphere by convection.
- Energy is transferred between the geosphere and the atmosphere by conduction.

Name ______________________________ Class ________________ Date ____________

Skills Worksheet

Vocabulary and Section Summary B

The Cycling of Matter

VOCABULARY

After you finish reading the section, try this puzzle! Look at the clues below, and then write the terms being described in the blanks on the next page. The boxed letters will spell out a new phrase; write the phrase in the space provided.

1. rocks that form when rock is changed because of chemical processes or changes in temperature and pressure
2. the process in which liquid water changes into gaseous water vapor
3. the process by which rock is broken down by wind, water, and temperature changes
4. any form of water that falls to Earth's surface from the clouds
5. the change from a gas to a liquid
6. rocks that form when rocks break into smaller pieces and those pieces become cemented together
7. the process in which nitrogen circulates among the air, soil, water, plants, and animals in an ecosystem
8. the process in which water vapor is released into the air through pores on the leaves of plants
9. the process by which wind, water, ice, or gravity transports soil and sediment from one location to another
10. the movement of carbon from the nonliving environment into living things and back
11. the continuous movement of water between the atmosphere, the land, and the oceans
12. the series of processes in which rock forms, changes from one type to another, is destroyed, and forms again by geologic processes
13. rocks that form when magma cools and becomes solid

1. ☐ _ _ ☐ _ _ _ _ _ _ _ _ _ _ _ _ _ _ _

2. _ _ _ _ _ _ _ ☐ _ _ _

3. _ _ _ ☐ _ ☐ _ _ _ _ _

4. _ ☐ _ _ _ _ _ _ _ _ ☐ _ _

5. _ _ _ _ _ _ _ ☐ _ _ _ _ _

6. _ _ _ _ _ _ _ _ ☐ _ ☐ _

7. _ _ _ _ _ _ _ _ _ _ _ _ _ _ _ _

8. _ _ ☐ _ _ _ _ _ _ _ _ _ _ ☐

9. _ _ _ ☐ _ _ _

10. _ _ _ _ _ _ _ _ _ _ ☐

11. _ _ _ _ ☐ _ _ _ _ _ _

12. ☐ _ _ _ _ _ _ _ _

13. _ _ _ ☐ _ _ _ _ _ _ _ _ _ _

14. What is the phrase?

_ _ _ _ _ _ _ _ _ _ _ _ _ _ F _ _ _ _ D

SECTION SUMMARY

Read the following section summary.

- The processes that cycle matter in the Earth system can be relatively rapid or may take millions of years.
- The rock cycle is the series of processes in which rock changes from one form to another by geologic processes.
- The three major classes of rocks are sedimentary, igneous, and metamorphic.
- Water moves continuously from the ocean, to the atmosphere, to land, and back to the ocean through the water cycle.
- In the carbon cycle, carbon is cycled in both rapid processes and slow processes.
- Types of matter that are cycled through the Earth system include carbon, phosphorus, and nitrogen.

Name ______________________ Class ______________ Date ______________

Skills Worksheet

Directed Reading B

Section: Natural Resources (pp. 128–131)

1. What do the water you drink, the paper you write on, and the air you breathe have in common?

____

____

EARTH'S RESOURCES

______ **2.** Any natural material that is used by humans is called a(n)

a. human resource.
b. Earth resource.
c. natural resource.
d. recyclable resource.

______ **3.** Which of the following is NOT a natural resource?

a. mineral
b. petroleum
c. forest
d. plastic

4. Where does the energy we get from many of our natural resources ultimately come from?

____

5. What is a renewable resource?

____

____

____

6. Give one example of a renewable resource.

____

7. What is a nonrenewable resource?

____

____

____

8. Give three examples of a nonrenewable resource.

____

____

____

CONSERVING NATURAL RESOURCES

9. What is one way in which people can conserve natural resources?

____

____

____

10. What is one reason lakes, rivers, and other water resources should be kept free of pollution?

____

____

11. Name two ways in which people can conserve energy.

____

____

12. Define *recycling*.

____

____

____

13. How does recycling help conserve energy?

____

____

____

____

14. Name three kinds of products that are recyclable.

____

____

____

Name ______________________ Class ______________ Date ______________

Skills Worksheet

Directed Reading B

Section: Rock and Mineral Resources (pp. 132–137)

1. A naturally formed solid that has a crystalline structure is a(n)

______________________.

2. Where do minerals form?

____

____

3. The natural material that makes up most of the solid part of Earth is

______________________.

Match the correct description with the correct term. Write the letter in the space provided.

_____ **4.** A body of water dries up.

_____ **5.** Magma rises upward and cools.

_____ **6.** Groundwater is heated by magma.

_____ **7.** Surface water and groundwater carry dissolved materials into lakes and seas.

_____ **8.** Changes take place in pressure, temperature, and chemical makeup.

a. metamorphism

b. cooling of plutons

c. deposition

d. reactions between hot water and rocks

e. evaporation

9. Name the five characteristics geologists use to identify a mineral.

____

____

____

____

____

____

MINING

10. Why are rocks and minerals mined from the ground?

____

____

11. A deposit of desired material large enough to be mined for profit is called

a(n) ______________________.

12. Name the two methods by which rocks and minerals are removed from the ground.

____

____

13. What kind of mines are open pits and quarries?

____

14. When coal is removed in strips, it is called ______________________.

15. What is used when ore is too deep within Earth to be surface mined?

____

16. What is dug into the ground in subsurface mining?

____

17. List two problems that can result from mining.

____

____

18. The process of returning land to its original state after mining is called

______________________.

MAKING COMMON OBJECTS

19. Name a mineral that can be used as it is.

____

20. Name three characteristics of metals.

____

____

____

21. Name three characteristics of nonmetals.

____

____

____

22. What is a major component of cement?

____

23. What is gypsum used to make?

____

____

____

Name ______________________ Class ______________ Date ____________

Skills Worksheet

Directed Reading B

Section: Using Material Resources (pp. 138–143)

1. Where do all of the objects that you need to live come from?

____

RESOURCES FROM EARTH

_____ **2.** Into what two groups can Earth's resources be divided?
- **a.** natural and artificial
- **b.** energy and material
- **c.** natural and material
- **d.** energy and organic

_____ **3.** What kind of resources do humans consume or use to make objects?
- **a.** material
- **b.** natural
- **c.** atmospheric
- **d.** energy

4. List the four sources on Earth for material resources.

____

____

____

____

5. What is the most valuable resource in the atmosphere?

____

Match the correct description with the correct term. Write the letter in the space provided.

_____ **6.** used as a source of chlorine

_____ **7.** used to help reduce car exhaust

_____ **8.** used inside light bulbs

_____ **9.** used to provide gasoline

_____ **10.** used to make steel

_____ **11.** used to make pipes for plumbing

a. petroleum
b. iron
c. polymer
d. argon
e. platinum
f. salt

12. A liquid mixture of complex hydrocarbons is called

____________________.

RESOURCES FROM LIVING THINGS

13. Why do humans harvest and consume plants?

____

____

14. Name three products that people use that are made from plant resources.

____

____

____

15. Name five products that people use that are made from animal resources.

____

____

____

____

____

THE COSTS OF MATERIAL RESOURCES

16. Name two examples of costs related to using plant and animal resources.

____

____

17. What is the basic requirement for commercial products?

____

____

18. What must be used if a material resource becomes too expensive to obtain?

____

____

19. Name two possible environmental costs of a product.

____

____

20. What are three things that some environmental protection laws require of a community?

____

____

____

Skills Worksheet

Vocabulary and Section Summary B

Natural Resources

VOCABULARY

After you finish reading the section, try this puzzle! In each statement below, the underlined term is incorrect. Write the correct term in the space provided.

1. Any natural material that is used by humans is a(n) <u>atmospheric resource</u>.

____

2. <u>Mining</u> is the process of reusing materials from waste or scrap.

____

3. An example of a(n) <u>renewable resource</u> is coal, which takes millions of years to form.

____

4. An example of a(n) <u>environmental resource</u> is fresh water, which can be replaced in a relatively short time.

____

5. Earth's atmosphere maintains air <u>composition</u> and produces rain.

____

SECTION SUMMARY

Read the following section summary.

- We use natural resources such as fresh water, petroleum, and trees to make our lives easier and more comfortable.
- Renewable resources can be replaced in a relatively short time, but nonrenewable resources may take thousands or even millions of years to form.
- Natural resources can be conserved by using only what is needed, by taking care of resources, and by reusing and recycling.

Name ______________________ Class ______________ Date ______________

Skills Worksheet

Vocabulary and Section Summary B

Rock and Mineral Resources

VOCABULARY

After you finish reading the section, try this puzzle! Use the clues below to unscramble the letters, and write the correct terms in the blanks. Then, write the corresponding letters in the numbered blanks to answer the question.

1. a natural material that makes up most of the solid part of Earth's surface: CKRO

2. a process by which land used for mining is returned to its original state: LMAATRIENOC

3. a natural, inorganic solid that has an orderly internal structure: EALNIMR

4. what surface coal mining is sometimes known as: PTSIR GNIIMN

5. a natural material whose concentration of economically valuable minerals is high enough to be mined profitably: REO

6. a process during which there are changes in pressure, temperature, or chemical makeup: MOPMSMEATRHI

1. ___ ___ ___ ___
(blank 2 = 5, blank 3 = 9)

2. ___ ___ ___ ___ ___ ___ ___ ___ ___ ___ ___
(blank 3 = 1, blank 6 = 4)

3. ___ ___ ___ ___ ___ ___ ___
(blank 1 = 3)

4. ___ ___ ___ ___ ___ ___ ___ ___ ___ ___ ___
(blank 2 = 10, blank 8 = 6)

5. ___ ___ ___
(blank 1 = 2, blank 3 = 8)

6. ___ ___ ___ ___ ___ ___ ___ ___ ___ ___ ___ ___
(blank 6 = 7, blank 11 = 11)

7. What do minerals and ores make?

___ ___ ___ ___ ___ ___
1 2 3 4 5 6

___ b j ___ ___ ___ ___
7 8 9 10 11

SECTION SUMMARY

Read the following section summary.

- A mineral is a naturally formed, inorganic solid that has a definite crystalline structure and a consistent chemical composition.
- Environments in which minerals form may be located at or near Earth's surface or deep below the surface.
- Two types of mining are surface mining and subsurface mining.
- Two ways to reduce the harmful effects of mining are through the reclamation of mined land and the recycling of mineral products.
- Both metals and nonmetals are used to make common objects.

Name ______________________ Class ______________ Date ______________

Skills Worksheet

Vocabulary and Section Summary B

Using Material Resources

VOCABULARY

After you finish reading the section, try this puzzle! Use the clues below to solve the crossword puzzle.

ACROSS

2. polymers made from chemicals

3. a natural resource that humans use to make objects

4. the process of recovering useful materials from waste

DOWN

1. a natural, usually inorganic solid that has a characteristic chemical composition

2. a resource used to provide gas

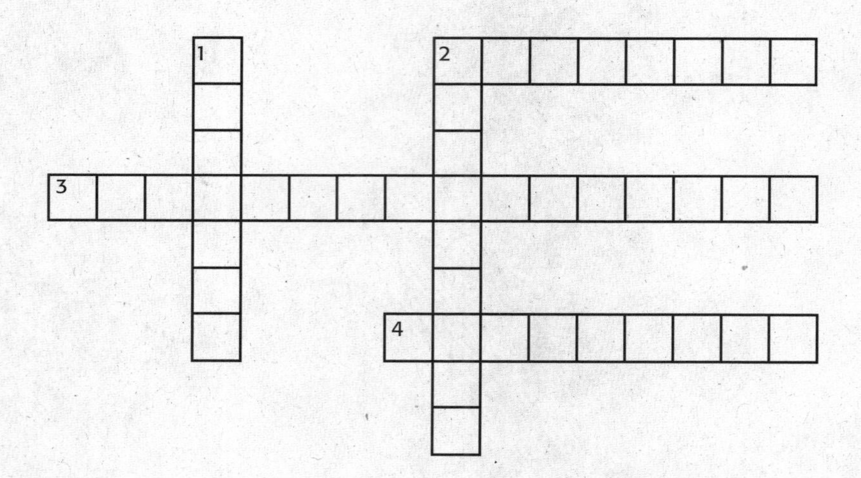

SECTION SUMMARY

Read the following section summary.

- Resources from Earth include gases from the atmosphere and rocks, minerals, and petroleum from Earth's crust.
- Living things provide humans with materials, such as food, clothing, and shelter.
- Using natural resources involves both economic and environmental costs.
- Reducing the environmental cost of using resources sometimes involves increasing the economic cost.

Name ______________________ Class ______________ Date ____________

Skills Worksheet

Directed Reading B

Section: Fossil Fuels (pp. 158–165)

______ **1.** Natural resources that humans use to generate energy are called
a. acid resources.
b. energy resources.
c. human resources.
d. Earth resources.

FOSSIL FUELS AS ENERGY RESOURCES

______ **2.** A nonrenewable energy resource formed from the remains of plants and animals that lived long ago is called a(n)
a. fossil fuel.
b. hydrogen fuel.
c. bio fuel.
d. clean fuel.

3. Give three examples of fossil fuels.

____

____

____

4. Why is it important to conserve fossil fuels?

____

____

____

____

TYPES OF FOSSIL FUELS

______ **5.** All living things contain the element
a. hydrogen.
b. carbon.
c. calcium.
d. iron.

______ **6.** Most carbon in fossil fuels exists as hydrogen-carbon compounds called
a. carbohydrates.
b. hydrocarbons.
c. petroleum distillates.
d. black gold.

______ **7.** A liquid mixture of complex hydrocarbon compounds is called
a. natural gas.
b. coal.
c. petroleum.
d. hydrogen.

______ **8.** What percentage of the world's energy comes from petroleum products?
a. only 10%
b. only 20%
c. less than 30%
d. more than 40%

______ **9.** A mixture of gaseous hydrocarbons used as fuel is called
a. natural gas.
b. coal.
c. petroleum.
d. hydrogen.

______ **10.** What is one disadvantage of using natural gas as fuel?
a. Natural gas causes more air pollution than oil does.
b. Natural gas is hard to burn.
c. Natural gas leaks can cause explosions.
d. Natural gas cannot be used to generate electricity.

______ **11.** What is the main component of natural gas?
a. butane
b. crude
c. propane
d. methane

______ **12.** A solid fossil fuel formed underground from partially decomposed plant material is called
a. natural gas.
b. coal.
c. petroleum.
d. hydrogen.

13. What are two ways that coal was used when it was the major source of energy in the United States?

____

____

14. What is one reason that people now use less coal?

____

____

15. What is one major way coal is used today?

____

____

HOW DO FOSSIL FUELS FORM?

16. How are all fossil fuels formed?

____

____

____

17. How do petroleum and natural gas form from microscopic sea organisms?

____

____

____

____

18. Rocks through which petroleum and gas can move are called

______________________ rocks.

19. What will happen to part of the remains of today's sea life millions of years from now?

____

____

____

Match the correct description with the correct term. Write the answer in the space provided.

______ **20.** brown, crumbly matter made of plant material, burned for heat or used as fuel

______ **21.** coal that is about 70% carbon, created when pressure and heat applied to peat force out water and gases,

______ **22.** coal that is about 80% carbon, created when sediment and pressure force water and gases out of lignite

______ **23.** the hardest type of coal, about 90% carbon

a. bituminous coal
b. peat
c. lignite
d. anthracite

24. The greater the ______________________ content of coal, the more cleanly the coal burns.

WHERE FOSSIL FUELS ARE FOUND

______ **25.** Approximately how much of the petroleum used by the United States is imported from other countries?

a. none
b. about one-fourth
c. over one-half
d. all

HOW FOSSIL FUELS ARE OBTAINED

______ **26.** How are petroleum and natural gas removed from Earth's crust?
- **a.** by drilling wells
- **b.** by deep mining
- **c.** by strip mining
- **d.** by surface mining

27. Give two names for the process by which soil and rock are stripped from Earth's surface to expose underlying coal.

____

____

PROBLEMS WITH FOSSIL FUELS

28. Rain, sleet, or snow that has a high concentration of acids, often because of air pollutants, is called ______________________.

29. What is one environmental problem caused by coal mining?

____

____

____

30. What is one problem caused by producing, transporting, and using petroleum?

____

____

____

31. The photochemical haze that forms when sunlight acts on industrial pollutants and burning fuels is called ______________________.

32. What are two reasons some large cities have serious smog problems?

____

____

____

Name ______________________ Class ______________ Date __________

Skills Worksheet

Directed Reading B

Section: Alternative Energy (pp. 166–173)

NUCLEAR ENERGY

______ **1.** The energy released by a fission or fusion reaction is
- **a.** wind energy.
- **b.** chemical energy.
- **c.** hydroelectric energy.
- **d.** nuclear energy.

______ **2.** Where does fusion occur naturally?
- **a.** in the sun
- **b.** in the moon
- **c.** in a cooling tower
- **d.** in a strip mine

3. The process by which the nuclei of radioactive atoms are split into two or more smaller nuclei is called ______________________.

4. What are two reasons people do not use nuclear energy instead of using fossil fuels?

____

____

____

____

5. The joining of two or more nuclei to form a larger nucleus is ______________________.

6. What is one of the main advantages of fusion?

____

____

7. What is the main disadvantage of fusion?

____

____

WIND ENERGY

______ **8.** The use of a windmill to drive an electric generator is
- **a.** nuclear energy.
- **b.** chemical energy.
- **c.** solar energy.
- **d.** wind power.

______ **9.** Which of the following is true about wind energy?
a. Wind energy is renewable.
b. Wind energy is created by fission.
c. Wind energy causes a lot of pollution.
d. Wind energy is created by fusion.

CHEMICAL ENERGY FROM FUEL CELLS

______ **10.** Fuel cells produce chemical energy by combining
a. carbon and oxygen.
b. hydrogen and oxygen.
c. carbon and hydrogen.
d. carbon, hydrogen, and oxygen.

______ **11.** What is the only byproduct of fuel cells?
a. smog
b. methane
c. radioactive waste
d. water

______ **12.** The energy released when a chemical compound reacts to produce new compounds is called
a. nuclear energy.
b. chemical energy.
c. solar energy.
d. wind power.

SOLAR ENERGY

______ **13.** The energy received by Earth from the sun in the form of radiation is
a. nuclear energy.
b. chemical energy.
c. solar energy.
d. wind power.

14. Is solar energy a renewable or nonrenewable resource?

____

15. What are solar cells or photovoltaic cells used for?

____

____

____

16. What is one advantage of solar energy?

____

____

____

17. What is one disadvantage of solar energy?

____

____

Name ______________________ Class ______________ Date ______________

HYDROELECTRIC ENERGY

______ **18.** Electrical energy produced by moving water is called
- **a.** wind power.
- **b.** hydraulic power.
- **c.** hydrogen energy.
- **d.** hydroelectric energy.

19. What is one advantage of using hydroelectric energy?

____

20. What is one disadvantage of using hydroelectric energy?

____

____

21. What are three problems that could be caused by building hydroelectric dams?

____

____

____

ENERGY FROM LIVING THINGS

______ **22.** Any organic matter that can be used as a source of energy is called
- **a.** gasohol.
- **b.** treated effluent.
- **c.** biomass.
- **d.** critical mass.

______ **23.** The most common way to release biomass energy is to
- **a.** soak biomass in water.
- **b.** mix biomass with coal.
- **c.** store biomass in tanks.
- **d.** burn biomass.

24. Plants that contain sugar or starch can be made into what product?

____

25. Alcohol can be mixed with gasoline to make a fuel called

______________________.

26. What is one advantage of using biomass as a source of energy?

____

27. What is one disadvantage of using biomass as a source of energy?

____

____

ENERGY FROM WITHIN EARTH

______ **28.** The energy produced by the heat within Earth is

a. solar energy.
b. biomass energy.
c. biochemical energy.
d. geothermal energy.

______ **29.** Melted rock is called

a. magma.
b. steam.
c. groundwater.
d. biomass.

30. Describe how geothermal energy can be used to heat a building.

____

____

____

____

Name ______________________ Class ______________ Date ____________

Skills Worksheet

Vocabulary and Section Summary B

Fossil Fuels

VOCABULARY

After you finish reading the section, try this puzzle! Fill in the blanks with the correct terms. Then, use the words to complete the word search puzzle on the next page.

1. A nonrenewable energy resource formed from the remains of organisms that lived long ago is a(n) ______________________.

2. A liquid mixture of complex hydrocarbon compounds is called ______________________.

3. A mixture of gaseous hydrocarbons located under Earth's surface is called ______________________.

4. Precipitation, such as rain, sleet, or snow, that contains a high concentration of acids is called ______________________.

5. A fossil fuel that forms underground from partially decomposed plant material is ______________________.

6. A(n) ______________________ is a natural resource that humans use to generate energy.

A	W	X	V	K	F	B	Z	B	N	I	R	Q	A	R	K	G	I	F	N
Y	C	C	Y	G	C	V	K	A	J	J	C	J	G	X	J	G	T	O	Y
Q	O	I	X	S	C	D	T	H	K	Z	C	B	O	J	N	B	R	S	R
Y	H	Z	D	Z	M	U	K	O	Y	X	Q	P	R	G	B	L	E	S	G
W	N	R	S	P	R	A	A	N	D	D	U	D	V	R	M	E	Z	I	J
C	R	G	R	A	R	F	O	G	W	C	W	U	C	J	N	C	T	L	W
L	T	W	L	D	T	E	D	E	L	R	H	B	Q	M	N	R	F	F	D
O	A	G	M	N	N	V	C	R	Z	A	G	S	B	X	F	U	G	U	L
L	A	O	O	Y	Q	I	P	I	S	U	Q	Y	E	G	X	O	Z	E	Q
S	S	U	C	J	C	P	J	X	P	Z	M	D	A	I	Y	S	T	L	G
Y	R	S	Y	S	X	U	V	B	B	I	S	S	V	V	S	E	M	G	N
V	P	Q	G	G	S	H	N	D	O	U	T	A	M	Y	Q	R	T	T	S
K	K	L	S	R	K	L	Z	D	H	D	J	A	Z	Y	T	Y	X	P	W
N	E	T	Z	T	T	T	D	F	V	G	G	Z	T	O	K	G	X	W	O
X	V	S	B	H	E	R	S	D	M	K	Q	C	R	I	W	R	H	E	U
U	E	M	R	P	O	Y	T	M	A	R	N	L	J	Y	O	E	O	Q	K
L	Y	D	D	F	A	A	N	P	S	Q	Z	K	T	N	A	N	X	E	K
M	H	B	Q	J	A	W	U	N	V	Z	E	K	I	J	O	E	I	M	A
M	U	E	L	O	R	T	E	P	K	I	P	J	K	L	H	L	V	J	F
X	D	U	C	A	P	Y	B	D	T	M	A	E	W	L	P	Z	A	R	C

SECTION SUMMARY

Read the following section summary.

- Energy resources are natural resources that humans use to produce energy.
- Fossil fuels are nonrenewable resources that form slowly over long periods of time from the remains of dead organisms. Petroleum, natural gas, and coal are fossil fuels.
- When fossil fuels are burned, they release energy. Most of that energy is heat energy.
- How humans use fossil fuels depends on the availability of the fuel, the ways in which the fuels are converted into energy, and the effects of converting the fuels into energy.
- Fossil fuels are found all over the world. The United States imports more than half of the petroleum that it uses from the Middle East, South America, Africa, Mexico, and Canada.
- Fossil fuels are obtained by drilling oil wells, mining below Earth's surface, and strip mining.
- Acid precipitation, smog, water pollution, and the destruction of wildlife habitats are some of the environmental problems created by the use of fossil fuels.

Name ______________________ Class ______________ Date ____________

Skills Worksheet

Vocabulary and Section Summary B

Alternative Energy

VOCABULARY

After you finish reading the section, try this puzzle! In each of the following items, use the clue to unscramble the letters, and write the term in the corresponding blanks.

1. fuel made from gasoline and alcohol: HOSLGOA

___ ___ ___ ___ ___ ___ ___
(1 under blank 3)

2. energy that comes from the sun: LSOAR EERNGY

___ ___ ___ ___ ___   ___ ___ ___ ___ ___ ___
(2 under blank 1)

3. energy released when a compound reacts to make a new compound: MCCHIEAL GNREEY

___ ___ ___ ___ ___ ___ ___ ___   ___ ___ ___ ___ ___ ___
(2 under the last blank of the first word; 2 under blank 5 of the second word)

4. created by falling water: DECHOLRIETYRC NREGEY

___ ___ ___ ___ ___ ___ ___ ___ ___ ___ ___ ___ ___   ___ ___ ___ ___ ___ ___
(1 under blank 1 of the second word)

5. the use of a windmill to run an electric generator: NWDI WPREO

___ ___ ___ ___   ___ ___ ___ ___ ___
(2 under blank 1)

6. organic matter that contains stored energy: OSBSIMA

___ ___ ___ ___ ___ ___ ___
(2 under blank 2)

7. released by fusion and fission: NCLUAER EYEGNR

___ ___ ___ ___ ___ ___ ___   ___ ___ ___ ___ ___ ___
(1 under blank 2)

8. produced by heat within Earth: EHRELGAMOT YENGRE

___ ___ ___ ___ ___ ___ ___ ___ ___ ___   ___ ___ ___ ___ ___ ___
(2 under blank 2)

Now go back through items 1–8. Group the numbered letters together by number, and unscramble each group to make a word. Then, write the unscrambled word for each group in the corresponding boxes to form a statement about our use of alternative energy resources.

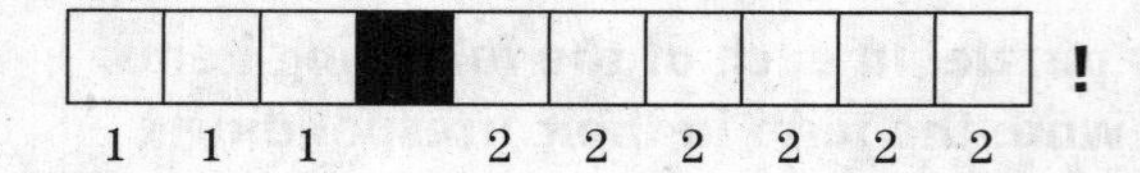

SECTION SUMMARY

Read the following section summary.

- The usefulness of a resource depends on how easy converting the resource into energy is, how expensive the resource is, and how much of the resource is available.
- Fission and fusion are processes that release nuclear energy. The byproducts of fission are heat and radioactive waste.
- Wind power, solar energy, hydroelectric energy, biomass, and geothermal energy are renewable resources that emit very little pollution.
- Not all alternative energy resources can be generated in all areas. Some alternative energy resources are very expensive.
- Every energy resource has advantages and disadvantages.

Name ______________________ Class ______________ Date ____________

Skills Worksheet

Directed Reading B

Section: Earth's Structure (pp. 190–197)

THE LAYERS OF EARTH

1. Earth is composed of several ______________

2. Scientists think about Earth's layers in terms of their

______________________ composition and their ______________________

properties.

3. List the three layers of Earth, based on their chemical composition.

____

____

____

4. What element makes up most of Earth's core?

____

5. Name the three elements that make up most of Earth's mantle.

____

____

____

6. What three elements make up most of Earth's crust?

____

____

____

7. How much of Earth's mass is made up by the core?

____

____

8. Oceanic crust is denser than the continental crust because it contains more of which three elements?

____

____

____

Match the correct description with the correct term. Write the letter in the space provided.

______ **9.** the solid, outer layer of Earth that consists of the crust and the rigid upper part of the mantle

______ **10.** the soft layer of the mantle on which the tectonic plates move

______ **11.** the liquid layer of the core

______ **12.** the solid layer of the core

______ **13.** the lower part of the mantle

a. asthenosphere
b. lithosphere
c. mesosphere
d. outer core
e. inner core

MAPPING EARTH'S INTERIOR

______ **14.** What do scientists use to study Earth's interior?
- **a.** global positioning systems
- **b.** tectonic plates
- **c.** oceanic waves
- **d.** seismic waves

______ **15.** What are seismic waves?
- **a.** movements in the outer core
- **b.** pictures of the Earth's interior
- **c.** vibrations from an earthquake
- **d.** vibrations from a seismometer

16. What affects the speed of seismic waves?

____

____

CONTINENTAL DRIFT

______ **17.** What hypothesis by Alfred Wegener explains why continents seem to fit together?
- **a.** continental spreading
- **b.** plate tectonics
- **c.** Wegener's puzzle
- **d.** continental drift

______ **18.** According to Wegener, how many landmasses did all continents once form?

a. one
b. six
c. seven
d. ten

______ **19.** What did Wegener hypothesize happened to the continents?

a. They broke up and re-formed.
b. They drifted together to form a single continent.
c. They broke up and drifted to their current locations.
d. They sank into the ocean.

20. Does fossil evidence support Wegener's theory? Explain your answer.

____

____

____

21. List three kinds of evidence found on continents that are far from each other that support Wegener's theory.

____

____

____

THE BREAKUP OF PANGAEA

22. Wegener thought that all the present continents were once joined 245 million years ago in a single large continent he called

______________________.

23. The single continent split into two continents called Gondwana and

______________________ about 135 million years ago.

24. What was formed when these two continents split again 65 million years ago?

____

____

SEA-FLOOR SPREADING

25. Why did many scientists reject Wegener's hypothesis?

____

____

Match the correct definition with the correct term. Write the letter in the space provided.

_____ **26.** process of forming new oceanic lithosphere as magma rises to Earth's surface

_____ **27.** underwater mountain chains that run through Earth's ocean floor

_____ **28.** process that happens when Earth's magnetic poles change places

_____ **29.** theory that explains how continents reached their current locations

a. continental drift
b. mid-ocean ridges
c. sea-floor spreading
d. magnetic reversal

30. It is now known that ______________________ is a mechanism by which continents move.

Name ______________________ Class ______________ Date ____________

Skills Worksheet

Directed Reading B

Section: The Theory of Plate Tectonics (pp. 198–203)

1. The theory that Earth's lithosphere is divided into tectonic plates that move around on top of the asthenosphere is called ______________________.

TECTONIC PLATES

2. Some tectonic plates contain both ______________________ crust and ______________________ crust.

3. Tectonic plates fit together like the pieces of a(n) ______________________.

4. How are tectonic plates like ice cubes in a bowl of punch?

____

____

____

TECTONIC PLATE BOUNDARIES

______ **5.** The place where tectonic plates touch is known as the

a. continental plate.
b. tectonic plate boundary.
c. magma zone.
d. tectonic ridge.

6. The boundary formed when tectonic plates collide is called a(n) ______________________ boundary.

7. List the three types of collisions that can occur at a convergent boundary.

____

____

____

8. When two plates made of continental lithosphere collide, ______________________ may form.

9. The process of denser crust sinking beneath less dense crust after collision is called ______________________.

10. At a(n) ______________________ boundary, two tectonic plates separate from each other.

11. Where do most divergent boundaries happen?

____

12. At what type of boundary do two tectonic plates slide past one another horizontally?

____

13. The San Andreas fault system in California is an example of what type of boundary?

____

CAUSES OF TECTONIC PLATE MOTION

______ **14.** When rock is heated, it becomes less dense and
- **a.** rises toward Earth's surface.
- **b.** sinks.
- **c.** moves sideways.
- **d.** erupts.

______ **15.** When rock cools, it tends to
- **a.** rise toward Earth's surface.
- **b.** sink below the surface.
- **c.** move sideways.
- **d.** push against the surface.

16. Density differences in the asthenosphere are caused by the flow of ______________________ within Earth.

Match the correct definition with the correct term. Write the letter in the space provided.

______ **17.** plate motion due to higher densities

______ **18.** plate motion due to gravity

______ **19.** plate motion due to the heating and cooling of rocks

a. ridge push

b. convection

c. slab pull

TRACKING TECTONIC PLATE MOTION

______ **20.** At what rate do tectonic plates move?
- **a.** kilometers per year
- **b.** meters per year
- **c.** meters per month
- **d.** centimeters per year

______ **21.** What do scientists use to measure the rate of tectonic plate movement on continents?
- **a.** clinometers
- **b.** global positioning system (GPS)
- **c.** densitometers
- **d.** seismometers

Name ______________________ Class ______________ Date ____________

Skills Worksheet

Directed Reading B

Section: Deforming Earth's Crust (pp. 204–209)

DEFORMATION

______ **1.** What is the amount of force per unit area on a given material called?
- **a.** bending
- **b.** stretching
- **c.** stress
- **d.** breakage

______ **2.** The process by which the shape of a rock changes because of stress is called
- **a.** seismology.
- **b.** elasticity.
- **c.** deformation.
- **d.** re-formation.

3. Name two ways in which rock layers can deform when stress is placed on them.

____

____

FOLDING

______ **4.** The bending of rock layers due to stress in Earth's crust is known as
- **a.** faulting.
- **b.** folding.
- **c.** divergence.
- **d.** convergence.

______ **5.** A fold in which the oldest rock layers are in the center of the fold is called a(n)
- **a.** syncline.
- **b.** hinge.
- **c.** anticline.
- **d.** limb.

______ **6.** A fold in which the youngest rock layers are in the center of the fold is called a(n)
- **a.** syncline.
- **b.** hinge.
- **c.** anticline.
- **d.** limb.

Match the correct definition with the correct term. Write the letter in the space provided.

______ **7.** a fold that appears to be lying on its side

______ **8.** a fold in which one limb is tilted beyond 90°

______ **9.** a fold in which one limb may dip more steeply than the other limb does

a. overturned fold
b. asymmetrical fold
c. recumbent fold

FAULTING

______ **10.** A break in a body of rock along which one block slides relative to another is called a
- **a.** wall.
- **b.** slide.
- **c.** fault.
- **d.** fold.

______ **11.** When extension pulls rocks apart, it creates a
- **a.** normal fault.
- **b.** fold.
- **c.** reverse fault.
- **d.** strike-slip fault.

______ **12.** When contraction pushes rocks together, it creates a
- **a.** normal fault.
- **b.** mid-ocean ridge.
- **c.** reverse fault.
- **d.** strike-slip fault.

______ **13.** When forces cause rock to break and slip parallel to Earth's surface, they create a
- **a.** normal fault.
- **b.** fold.
- **c.** reverse fault.
- **d.** strike-slip fault.

14. When a fault is not vertical, a hanging wall and a(n) ______________________ are formed.

15. The hanging wall moves down relative to the footwall in a(n) ______________________ fault.

16. The hanging wall moves up relative to the footwall in a(n) ______________________ fault.

17. Name three ways of recognizing fault offset.

____

____

____

PLATE TECTONICS AND MOUNTAIN BUILDING

_____ **18.** Over long periods of time, the movement of tectonic plates around Earth's surface can cause which of the following to occur?
a. volcanoes
b. transform boundaries
c. mountain building
d. divergent boundaries

_____ **19.** What kind of mountain range is formed when rock layers are squeezed together and pushed upward?
a. folded mountains
b. fault-block mountains
c. volcanic mountains
d. strike-slip mountains

_____ **20.** What kind of mountain range is formed when tension causes large blocks of Earth's crust to drop down relative to other blocks?
a. folded mountains
b. fault-block mountains
c. volcanic mountains
d. strike-slip mountains

_____ **21.** What kind of mountain is formed when molten rock rises to the surface and erupts?
a. folded mountains
b. fault-block mountains
c. volcanic mountains
d. strike-slip mountains

Match the correct description with the correct term. Write the letter in the space provided.

_____ **22.** Appalachian Mountains

_____ **23.** Tetons

_____ **24.** Ring of Fire

a. volcanic mountains
b. folded mountains
c. fault-block mountains

Name ______________________ Class ______________ Date ______________

Skills Worksheet

Directed Reading B

Section: California Geology (pp. 210–217)

1. California has been at an active ______________________ for the past 225 million years.

2. The most important force in shaping California's geologic history has been ______________________.

BUILDING CALIFORNIA BY PLATE TECTONICS

______ **3.** Where was North America's western edge about 225 million years ago, compared to today?
- **a.** underwater
- **b.** on another plate
- **c.** farther east
- **d.** farther west

4. When Pangaea broke up, North America's western edge became an active ______________________ plate boundary.

5. How long ago did the most important period of geologic "building" take place in California?

____

6. List the three major tectonic plates influencing California's geologic history.

____

____

____

Match the correct description with the correct term. Write the letter in the space provided.

______ **7.** plate that completely subducted at one part of the boundary about 25 million years ago

______ **8.** what was created when the Pacific plate touched North America for the first time

______ **9.** remains from the ancient Farallon plate

______ **10.** plate that moved closer and closer to North America

a. Farallon plate
b. Pacific plate
c. Juan de Fuca plate
d. transform boundary

SUBDUCTION AND VOLCANISM

11. What were two results of the subduction of the Farallon plate?

____

____

12. Magma from the subduction of the Farallon plate formed a mass of granite known as the ______________________.

13. What is a batholith?

____

____

14. The area in which the Juan de Fuca plate is subducting under the North American plate is called the ______________________.

15. List one effect of the subduction of the Juan de Fuca plate.

____

SUBDUCTION AND ACCRETION

16. The process of accretion forms ______________________.

17. Name one mountain range in California that was formed through accretion.

____

18. Chunks of subducting plates that build the edges of continents are called ______________________.

19. In the foothills along the western side of the Sierra Nevadas are rocks that contain most of California's ______________________.

THE SAN ANDREAS FAULT SYSTEM

Match the correct description with the correct term. Write the letter in the space provided.

______ **20.** approximate separation between the North American and Pacific plates

______ **21.** a large depression in southern California bordered by active faults

______ **22.** the most famous transform plate boundary

______ **23.** approximate length of the San Andreas fault system

a. Los Angeles Basin
b. San Andreas fault system
c. 1,000 km
d. 315 km

PLATE TECTONICS AND THE CALIFORNIA LANDSCAPE

______ **24.** What force helped form much of California's landscape?
a. ocean water
b. gravity
c. fault lines
d. plate tectonics

______ **25.** What has helped form central and northern California's steep, rocky coastline?
a. gravity
b. fault lines
c. uplift
d. ocean water

Name ______________________ Class ______________ Date ______________

Skills Worksheet

Vocabulary and Section Summary B

Earth's Structure

VOCABULARY

After you finish reading the section, try this puzzle! Use the clues to help you unscramble each word below. Write your answer in the spaces provided.

1. the tectonic process that takes place along mid-ocean ridges: AES RFOOL REGIDANSP

____

2. the layer of Earth that is made mostly of iron: RCEO

____

3. the layer of solid rock that flows very slowly: SEEONHTAPSRHE

____

4. the layer of Earth made of the crust and the mantle: LHETPHESROI

____

5. the layer of Earth that has the most mass: METNAL

____

6. the theory that continents move apart from each other: TCOITAENLNN FDTRI

____

7. the thin and solid outermost layer of Earth above the mantle: RCSUT

____

SECTION SUMMARY

Read the following section summary

- Earth is made up of three layers—the crust, the mantle, and the core—based on chemical composition. Of these three layers, the core is made up of the densest materials. The crust and mantle are made up of materials that are less dense than the core.
- Earth is made up of five layers—the lithosphere, the asthenosphere, the mesosphere, the outer core, and the inner core—based on physical properties.
- Knowledge about the layers of Earth comes from the study of seismic waves caused by earthquakes.
- Wegener hypothesized that continents drift apart from one another now and that they have drifted in the past.
- Magnetic reversals that occur over time are recorded in the magnetic pattern of the oceanic crust, which provides evidence of sea-floor spreading and continental drift.
- Sea-floor spreading is the process by which new sea floor forms at mid-ocean ridges.

Name ______________________ Class ______________ Date ______________

Skills Worksheet

Vocabulary and Section Summary B

The Theory of Plate Tectonics

VOCABULARY

After you finish reading the section, try this puzzle! First, number the letters of the alphabet in the squares provided under each letter. The first two letters are done for you. Next, fill in the appropriate letters in the numbered spaces for each of the terms below. Then, provide a brief definition for each term in the space provided.

A	B	C	D	E	F	G	H	I	J	K	L	M	N	O	P	Q	R	S	T	U	V	W	X	Y	Z
1	2																								

1. ___ ___ ___ ___ ___ ___ ___ ___ ___   ___ ___ ___ ___ ___ ___ ___ ___
20 18 1 14 19 6 15 18 13   2 15 21 14 4 1 18 25

____

____

____

2. ___ ___ ___ ___ ___ ___   ___ ___ ___
9 19 12 1 14 4   1 18 3

____

____

____

3. ___ ___ ___ ___ ___ ___ ___ ___ ___ ___   ___ ___ ___ ___ ___ ___ ___ ___
3 15 14 22 5 3 20 9 15 14   3 21 18 18 5 14 20 19

____

____

____

4. ___ ___ ___ ___ ___ ___ ___ ___ ___ ___   ___ ___ ___ ___ ___ ___ ___ ___
3 15 14 22 5 18 7 5 14 20   2 15 21 14 4 1 18 25

____

____

____

5. ___ ___ ___ ___ ___ ___ ___ ___   ___ ___ ___ ___ ___
20 5 3 20 15 14 9 3   16 12 1 20 5

____

____

____

6. ___ ___ ___ ___ ___ ___ ___ ___ ___ ___
19 21 2 4 21 3 20 9 15 14

____

____

____

7. ___ ___ ___ ___ ___ ___ ___ ___ ___   ___ ___ ___ ___ ___ ___ ___ ___
4 9 22 5 18 7 5 14 20   2 15 21 14 4 1 18 25

____

____

____

8. ___ ___ ___ ___ ___   ___ ___ ___ ___ ___ ___ ___ ___ ___
16 12 1 20 5   20 5 3 20 15 14 9 3 19

____

____

____

SECTION SUMMARY

Read the following section summary

- Plate tectonics is the theory that explains how pieces of Earth's lithosphere move and change shape.
- Tectonic plates are large pieces of the lithosphere that move around on top of the asthenosphere.
- Boundaries between tectonic plates are classified as convergent, divergent, or transform.
- Convection is the main driving force of plate tectonics.
- Tectonic plates move a few centimeters per year. Scientists measure this rate by using GPS or by using sea-floor spreading.

Name ______________________ Class ______________ Date ______________

Skills Worksheet

Vocabulary and Section Summary B

Deforming Earth's Crust

VOCABULARY

After you finish reading the section, try this puzzle! Use the clues below to solve the crossword puzzle.

ACROSS

2. amount of force per unit on a given material

5. polished surfaces

6. break in a body of rock

7. fold in which the youngest rock layers are in the center

8. bending of rock layers due to stress

9. fold in which the oldest rock layers are in the center

DOWN

1. process by which the shape of a rock changes in response to stress

3. block of rock that lies below the plane of the fault

4. not symmetrical

7. row of cliffs formed by faulting

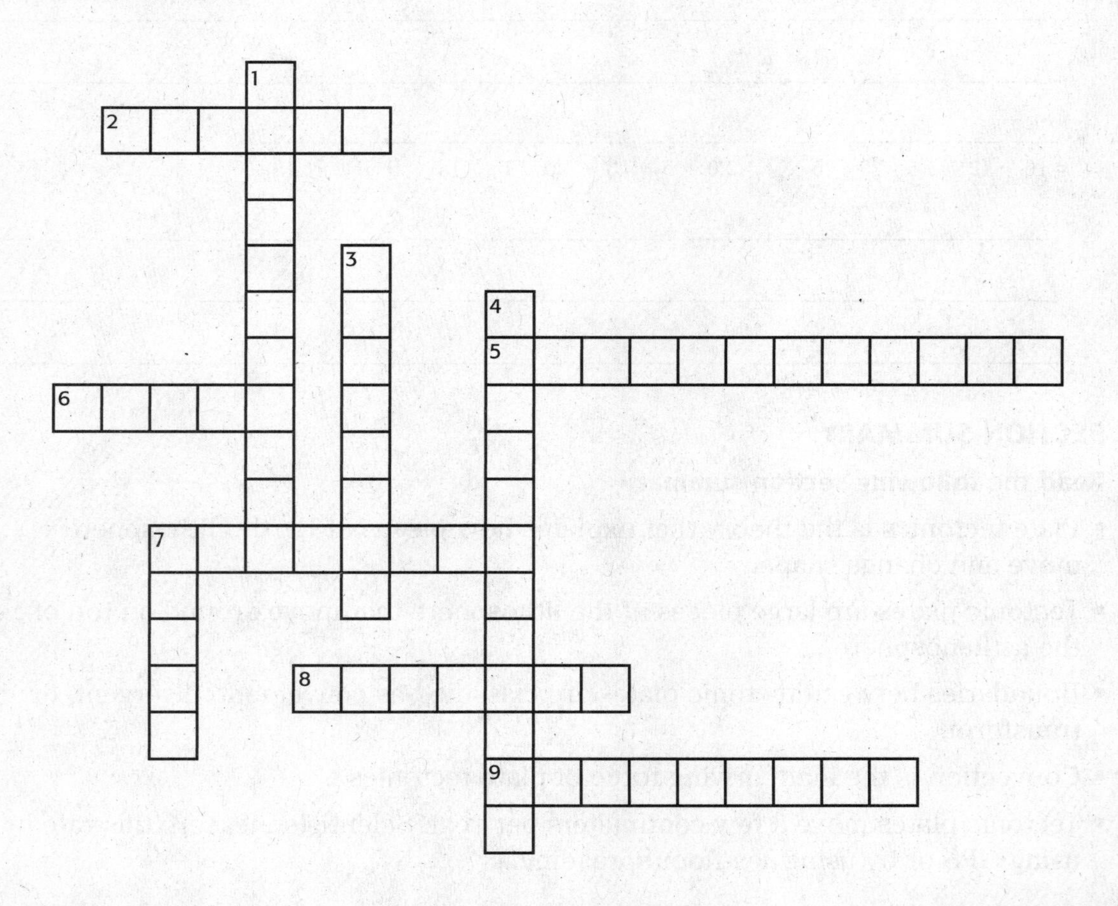

SECTION SUMMARY

Read the following section summary

- Deformation structures, such as faults and folds, form as a result of stress in the lithosphere. This stress is caused by tectonic plate motion.
- Folding occurs when rock layers bend because of stress.
- Faulting occurs when rock layers break because of stress and then move on either side of the break.
- Three major fault types are normal faults, reverse faults, and strike-slip faults.
- Mountain building is caused by the movement of tectonic plates. Folded mountains and volcanic mountains form at convergent boundaries. Fault-block mountains form at divergent boundaries.

Name ______________________ Class ______________ Date ______________

Skills Worksheet

Vocabulary and Section Summary B

California Geology

VOCABULARY

After you finish reading the section, try this puzzle! Use the clues below to complete the sentences, then find the terms in the word search puzzle on the following page. Words may be hidden horizontally, vertically, diagonally, or backward.

1. During the process called ______________, pieces of the plate that subducts may be scraped off and attached to the overriding plate, forming mountain chains parallel to the plate boundary.

2. A large mass of igneous rock that forms deep below the surface is called a(n) ______________.

3. The area off the northern coast of California where the Juan de Fuca plate is subducting beneath the North American plate is called the ______________.

4. A piece of lithosphere that becomes part of a larger landmass when tectonic plates collide at a convergent boundary is called a(n) ______________.

5. A mass of granite known as the ______________ was formed over 100 million years ago when a great deal of magma formed and solidified in the lithosphere as a result of the subduction of the Farallon plate.

SECTION SUMMARY

Read the following section summary

- Plate tectonics has been the most important force in the shaping of California's geology.
- When Pangaea broke apart, the western edge of North America became an active plate boundary.
- Between 225 million and 25 million years ago, subduction took place along all of California.
- During subduction, California grew larger as accreted terranes were added to the North American continent.
- A transform boundary formed about 25 million years ago, when the Pacific plate met the North American plate.
- Along the San Andreas fault, the Pacific plate is moving northwest relative to the North American plate.
- The motions of tectonic plates have caused mountains and valleys to form in California.

Name ______________________ Class ______________ Date ____________

Vocabulary and Section Summary B *continued*

C	C	W	P	B	S	R	L	Y	Y	N	R	T	J	C	C	N	X	B	V	G	S
N	A	A	Y	D	H	S	D	E	U	T	O	N	R	M	E	E	G	J	D	A	I
O	M	S	A	I	V	M	T	B	Y	H	C	I	E	Q	P	K	S	Q	C	V	E
A	F	M	C	L	E	C	C	H	T	D	A	P	T	T	O	S	V	C	J	B	R
D	Y	F	H	A	N	Z	R	N	X	Q	T	M	M	E	T	B	R	E	I	J	R
C	W	J	H	N	D	H	T	I	L	O	H	T	A	B	R	E	F	P	S	W	A
N	Y	L	U	Q	E	I	I	W	R	O	V	Q	S	N	T	C	Z	C	S	P	N
O	B	Z	Q	J	E	L	A	F	X	V	Q	K	K	E	Q	P	C	P	T	X	E
D	Z	Y	P	J	Q	X	O	S	N	D	E	Z	D	H	Q	H	O	A	W	Z	V
A	G	I	C	J	M	L	R	B	U	H	G	T	S	Y	C	A	K	D	W	X	A
G	A	B	Z	I	F	M	D	X	S	B	E	L	Q	F	R	J	T	N	X	H	D
Q	Z	X	F	H	P	W	F	Y	Y	R	D	C	C	B	P	M	I	H	J	A	A
L	B	U	T	M	M	L	L	M	R	G	V	U	F	P	F	K	J	Z	V	R	B
T	K	W	X	G	T	T	U	A	G	Y	R	N	C	V	F	G	Z	W	E	L	A
H	L	W	K	R	R	H	N	B	H	N	U	L	Q	T	E	H	V	Y	G	D	T
L	D	T	Q	C	R	E	V	Q	Q	V	D	N	U	I	I	E	C	E	Z	B	H
K	U	T	J	U	F	B	M	T	L	E	Z	N	H	F	J	O	E	H	X	Y	O
A	J	Y	B	G	D	O	Q	M	S	F	U	U	B	T	S	Q	N	S	L	C	L
M	L	Q	J	G	N	O	O	I	V	F	A	R	P	T	Z	E	F	Z	Z	S	I
I	D	Z	G	D	O	J	N	G	Q	I	L	H	L	H	I	N	G	Q	O	R	T
W	L	Q	B	V	M	W	J	C	Y	S	G	X	R	U	W	I	Y	P	S	N	H
B	J	E	X	Y	D	G	T	S	P	Z	D	P	T	E	W	V	F	U	O	S	E

Name ______________________ Class ______________ Date ____________

Skills Worksheet

Directed Reading B

Section: What Are Earthquakes? (pp. 232–237)

1. Define the term *earthquake.*

____

____

WHERE EARTHQUAKES HAPPEN

2. Where do most earthquakes take place?

____

____

3. Name two places where large earthquakes have occurred far from tectonic plate boundaries.

____

____

4. What is a fault?

____

____

FAULTS AT TECTONIC PLATE BOUNDARIES

Match the correct description with the correct term. Write the letter in the space provided.

_____ **5.** forms at divergent boundaries

_____ **6.** forms at convergent boundaries

_____ **7.** forms at transform boundaries

_____ **8.** forms where large numbers of interconnected faults are located

a. fault zone

b. normal fault

c. reverse fault

d. strike-slip fault

WHY EARTHQUAKES HAPPEN

Match the correct description with the correct term. Write the letter in the space provided.

______ **9.** a process that causes rock to change shape like a piece of clay being molded

______ **10.** a process that causes rock to change shape like a rubber band being stretched; leads to earthquakes

______ **11.** the sudden return of elastically deformed rock to its undeformed shape

______ **12.** a form of energy that travels through Earth, away from an earthquake in all directions

a. seismic waves
b. plastic deformation
c. elastic deformation
d. elastic rebound

13. When does elastic rebound happen?

____

____

____

14. Earthquake energy travels through rock as ______________________.

EARTHQUAKE WAVES

______ **15.** Seismic waves that travel through Earth's interior are called
a. surface waves.
b. body waves.
c. Earth waves.
d. tidal waves.

______ **16.** Seismic waves that travel along Earth's surface are called
a. surface waves.
b. body waves.
c. Earth waves.
d. primary waves.

______ **17.** Which of the following waves is a type of body wave?
a. surface wave
b. S wave
c. R wave
d. convergent wave

______ **18.** Which of the following waves is the fastest type of seismic wave?
a. surface wave
b. shear wave
c. S wave
d. P wave

______ **19.** Pressure waves are also known as
a. shear waves.
b. surface waves.
c. primary waves.
d. secondary waves.

______ **20.** Which of the following waves can travel through solids, liquids, and gases?
a. surface wave
b. S wave
c. P wave
d. shear wave

______ **21.** Which of the following is another name for *S wave?*
a. secondary wave
b. P wave
c. surface wave
d. primary wave

______ **22.** An S wave is unable to travel through
a. clay.
b. liquids.
c. sediments.
d. rock.

______ **23.** How many types of surface waves are there?
a. six
b. two
c. three
d. four

Match the correct description with the correct term. Write the letter in the space provided.

______ **24.** squeezes and stretches rock

______ **25.** shears rock horizontally from side to side

______ **26.** moves the ground with a rolling, up-and-down motion

a. S wave
b. surface wave
c. P wave

27. What does the speed of a seismic wave depend on?

____

____

28. Which type of earthquake wave is always the first to be detected?

____

29. Which type of body wave always arrives after P waves?

____

30. How do surface waves differ from body waves?

____

____

____

Name ______________________ Class ______________ Date ______________

Skills Worksheet

Directed Reading B

Section: Earthquake Measurement (pp. 238–243)

STUDYING EARTHQUAKES

Match the correct description with the correct term. Write the letter in the space provided.

_____ **1.** a tracing of earthquake motion

_____ **2.** the point on Earth's surface directly above an earthquake's starting point

_____ **3.** the location within Earth along a fault at which the first motion of an earthquake occurs

a. epicenter
b. seismogram
c. focus

4. Name two uses for seismometers.

____

____

____

Match the correct definition with the correct term. Write the letter in the space provided.

_____ **5.** the time between the arrival of P waves and S waves, which tells scientists how far the waves traveled

_____ **6.** the process by which circles are drawn around at least three seismometer stations, and the point of intersection is used to identify the epicenter

a. triangulation
b. lag time

EARTHQUAKE MAGNITUDE

_____ **7.** Seismograms provide information about earthquake
a. hazard.
b. depth.
c. damage.
d. strength.

Match the correct description with the correct term. Write the letter in the space provided.

______ **8.** a measure of earthquake strength

______ **9.** a way to determine magnitude by measuring ground motion from an earthquake and adjusting for distance

______ **10.** a way to determine magnitude by measuring the size of the fault area that moves, average distance that fault blocks move, and rock rigidity in the fault zone

a. moment magnitude scale

b. magnitude

c. Richter scale

EARTHQUAKE INTENSITY

______ **11.** The measurement of the effects of an earthquake at Earth's surface is called

a. magnitude.
b. intensity.
c. velocity.
d. seismology.

______ **12.** What intensity level on the Modified Mercalli scale indicates total destruction?

a. I
b. V
c. X
d. XII

______ **13.** The lowest intensity level on the Modified Mercalli scale describes an earthquake

a. that is felt by very few people.
b. that is not detected by seismometers.
c. that causes considerable damage.
d. that has a magnitude of 6.5 or greater.

14. Where are intensity values of an earthquake usually higher?

____

____

15. What would scientists use data from the Loma Prieta earthquake and the 1906 San Francisco earthquake for?

____

____

THE EFFECTS OF EARTHQUAKES

16. Name four factors that determine the effects of an earthquake on a given area.

____

____

____

____

17. What happens to the total energy in a seismic wave as the wave grows increasingly larger?

____

____

18. How does an area's distance from the epicenter affect the amount of damage an earthquake will do?

____

____

19. Name two effects of liquefaction.

____

____

20. Why are structures made of wood or steel more likely to survive strong ground shaking than structures made of brick or concrete?

____

____

Name ______________________ Class ______________ Date ____________

Skills Worksheet

Directed Reading B

Section: Earthquakes and Society (pp. 244–251)

EARTHQUAKE HAZARD

1. Define *earthquake hazard.*

____

____

____

2. Name one state that has a very high earthquake hazard level.

____

EARTHQUAKE FORECASTING

3. Strong earthquakes happen with less ______________ than weak earthquakes.

4. What is the relationship between earthquake strength and frequency based on?

____

____

____

____

5. What does the gap hypothesis state?

____

____

6. An area along a fault where relatively few earthquakes have taken place recently but where strong earthquakes have occurred in the past is called a(n)

______________________.

7. Some scientists believe that the gap hypothesis helped forecast the approximate location and strength of which earthquake?

____

REDUCING EARTHQUAKE DAMAGE

______ **8.** The process of making older buildings more earthquake resistant is called
a. quake proofing.
b. shock insulating.
c. mass dampening.
d. retrofitting.

______ **9.** In an earthquake-resistant building, which of the following structures act as shock absorbers?
a. mass dampers
b. base isolators
c. active tendon systems
d. cross braces

______ **10.** A weight placed in the roof of an earthquake-resistant building is called a(n)
a. mass damper.
b. base isolator.
c. active tendon system.
d. cross brace.

______ **11.** Which of the following structures are placed between floors to counteract the pressure on the sides of buildings during earthquakes?
a. flexible pipes
b. steel cross braces
c. mass dampers
d. base isolators

12. In an earthquake-resistant building, ______________________ help prevent waterlines and gas lines from breaking.

13. How does an active tendon system work in an earthquake-resistant building?

____

____

____

ARE YOU PREPARED FOR AN EARTHQUAKE?

14. Name two things you can do before an earthquake to make your home safer.

____

____

15. Name four supplies that would be helpful to have at home in case waterlines, power lines, gas lines, and roadways are damaged during an earthquake.

____

____

____

____

16. What should you do if you are indoors when an earthquake strikes?

____

____

17. What should you do if you are outside when an earthquake strikes?

____

____

18. Name three immediate dangers that people need to watch for after an earthquake ends.

____

____

____

TSUNAMIS

19. Define *tsunami*.

____

____

20. Why do tsunamis cause so much destruction?

____

____

21. Where are most of the nations that monitor tsunamis located?

____

____

Name ______________________ Class ______________ Date ____________

Skills Worksheet

Vocabulary and Section Summary B

What Are Earthquakes?

VOCABULARY

After you finish reading the section, try this puzzle! Use the clues below to solve the crossword puzzle.

ACROSS

3. a break in Earth's crust along which blocks of rock slide relative to one another

4. a wave of energy that travels through the Earth and away from an earthquake in all directions

5. a seismic wave that travels through Earth's interior

7. the process of one plate moving beneath another

8. a seismic wave that travels along Earth's surface

DOWN

1. a process by which rock deforms like a rubber band being stretched

2. the sudden return of elastically deformed rock to its undeformed shape

6. a movement or trembling of the ground that is caused by a sudden release of energy when rocks along a fault move

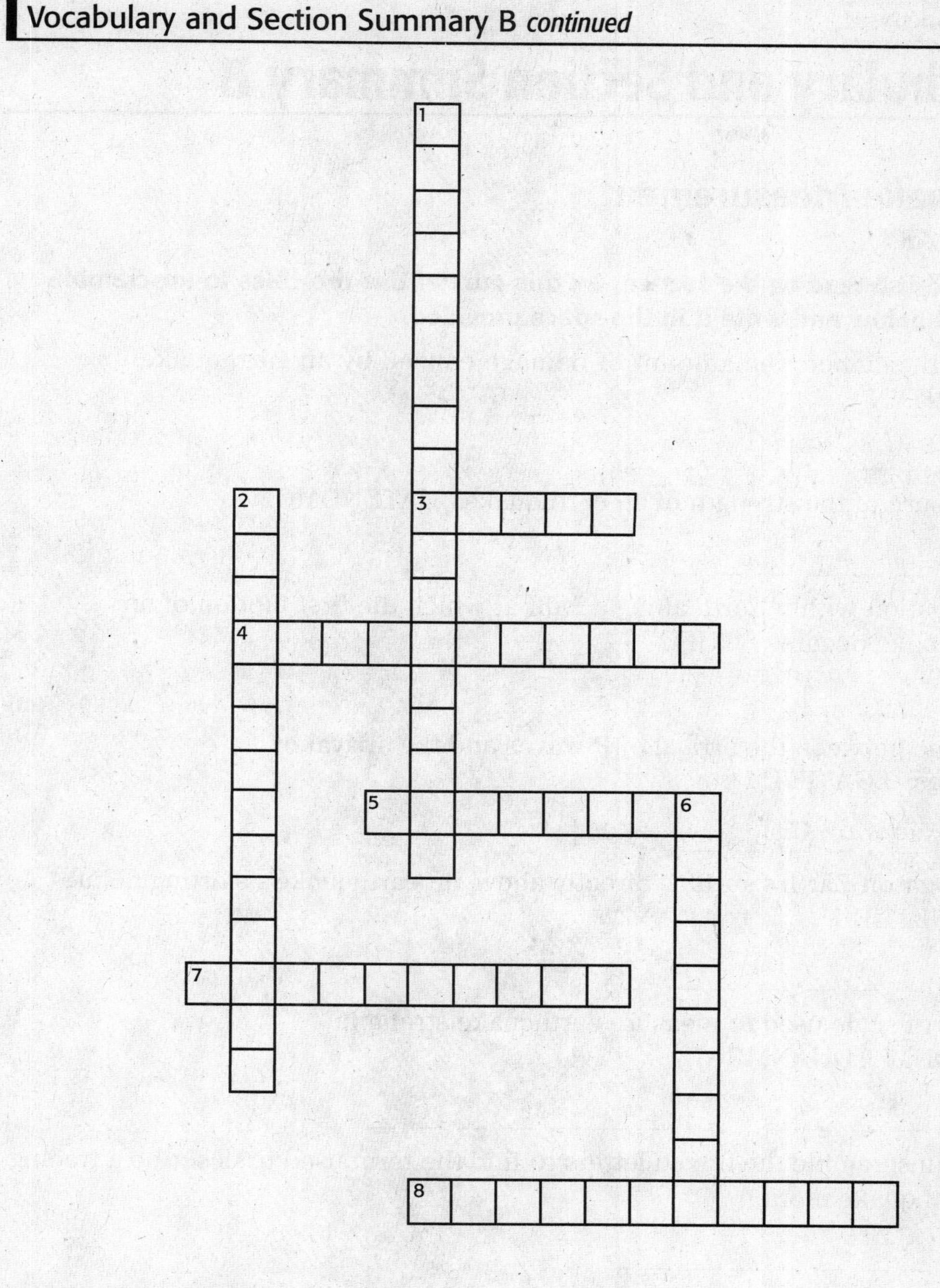

SECTION SUMMARY

Read the following section summary.

- Earthquakes are motions of the ground that happen as energy travels through rock.
- Earthquakes occur mainly near the edges of tectonic plates.
- Earthquakes are caused by elastic rebound, which is caused by sudden motions along faults. During elastic rebound, rock springs into its original shape and size as stress is released.
- Energy generated by earthquakes travels as body waves through Earth's interior or as surface waves along the surface of Earth.

Name ________________________________ Class ________________ Date ______________

Skills Worksheet

Vocabulary and Section Summary B

Earthquake Measurement

VOCABULARY

After you finish reading the section, try this puzzle! Use the clues to unscramble each word below and write it in the space provided.

1. in Earth science, the amount of damage caused by an earthquake: TESIINTYN

_ _ _ _ _ ☐ _ _ _

2. a measure of the strength of an earthquake: UATENDMGI

☐ ☐ _ _ _ _ _ _ ☐

3. the location within Earth along a fault at which the first motion of an earthquake occurs: SOUFC

_ ☐ _ _ ☐

4. the time between the arrival of P waves and the arrival of S waves: LGA IETM

_ _ ☐ _ ☐ _ _

5. the point on Earth's surface directly above an earthquake's starting point: TEEINRPCE

_ _ _ _ _ _ _ _ ☐

6. a type of scale used to measure earthquake strength: EMTONM TDUANMEGI

_ _ _ _ _ _ ☐ _ _ _ _ _ _ _ _

7. Now, unscramble the boxed letters to find the term used to describe a tracing of earthquake motion.

_ _ _ _ _ _ _ _ _ _

SECTION SUMMARY

Read the following section summary.

- An epicenter is the point on Earth's surface directly above where an earthquake started.
- The distance from a seismometer to an epicenter can be determined by using the lag time between P waves and S waves.
- An earthquake's epicenter can be located by triangulation.
- Magnitude is a measure of an earthquake's strength.
- Intensity is the effects of an earthquake.
- Important factors that determine the effects of an earthquake on a given area are magnitude, distance from the epicenter, local geology, and the type of construction.

Name ______________________ Class ______________ Date ____________

Skills Worksheet

Vocabulary and Section Summary B

Earthquakes and Society

VOCABULARY

After you finish reading the section, try this puzzle! Fill in the blanks in the clues below.

______________________ **1.** the basis for forecasting an earthquake's strength, location, and frequency; states that sections of active faults that have had few earthquakes recently are likely to be the sites of strong earthquakes in the future

______________________ **2.** the process of making older structures more resistant to earthquakes

______________________ **3.** an area along a fault where relatively few earthquakes have occurred recently but where strong earthquakes have occurred in the past

______________________ **4.** a giant ocean wave that forms after a volcanic eruption, submarine earthquake, or landslide

______________________ **5.** a measurement of how likely an area is to have damaging earthquakes in the future

Now see if you can find the vocabulary terms in the word search puzzle on the next page. Terms can be hidden in the puzzle vertically, horizontally, diagonally, or backward.

E	A	R	T	H	Q	U	A	K	E	H	A	Z	A	R	D	Q	W	T	U
G	U	L	H	S	A	W	B	Q	T	H	Y	K	I	E	K	L	N	S	R
J	A	D	V	F	R	T	O	Q	V	H	A	H	B	T	G	N	U	X	Y
B	P	P	H	R	L	A	O	X	T	X	X	C	R	R	Z	A	B	A	C
T	W	B	H	O	U	Z	J	A	M	S	C	K	T	O	V	N	J	W	E
B	G	S	L	Y	B	E	L	O	E	U	Z	J	A	F	U	U	P	V	F
D	Z	Q	F	G	P	L	A	I	Y	O	I	I	O	I	K	B	K	N	D
H	V	H	G	X	E	O	S	Z	D	O	V	R	N	T	I	S	L	V	T
Q	F	E	A	L	M	M	T	Y	I	S	G	T	Q	T	W	R	E	S	A
I	D	F	N	F	I	O	C	H	P	M	W	P	E	I	F	F	U	C	D
Q	D	L	H	C	Q	R	G	U	E	Y	Z	H	E	N	J	W	I	J	J
S	L	B	G	H	A	R	F	H	V	S	A	D	G	G	Z	H	B	F	E
X	C	A	R	I	F	X	M	M	L	S	I	U	P	Z	M	L	X	Q	F
J	P	D	I	G	I	L	Y	X	B	H	N	S	F	L	O	X	N	B	N
A	C	T	S	A	M	C	D	I	A	R	Z	G	I	M	A	N	U	S	T

SECTION SUMMARY

Read the following section summary.

- Earthquakes and tsunamis can affect human societies.
- Earthquake hazard is a measure of how likely an area is to have earthquakes in the future.
- Scientists use their knowledge of the relationship between earthquake strength and frequency and of the gap hypothesis to forecast earthquakes.
- Homes, buildings, and bridges can be strengthened to decrease earthquake damage.
- People who live in earthquake zones should safeguard their homes against earthquakes and have an earthquake emergency plan.
- Tsunamis are giant ocean waves that may be caused by earthquakes on the sea floor.

Name ______________________ Class ______________ Date ____________

Skills Worksheet

Directed Reading B

Section: Why Volcanoes Form (pp. 266–269)

1. A vent in Earth's surface through which magma and gases pass is called

a(n) ____________________.

WHERE VOLCANOES FORM

_____ **2.** Many volcanoes form
- **a.** near the center of continents.
- **b.** along bodies of water.
- **c.** along tectonic plate boundaries.
- **d.** in mountainous areas.

_____ **3.** What happens when magma reaches Earth's surface and erupts?
- **a.** Tectonic plates separate.
- **b.** An earthquake occurs.
- **c.** A rainbow forms.
- **d.** A volcano forms.

4. Why are the plate boundaries surrounding the Pacific Ocean called the Ring of Fire?

____

____

5. Liquid rock produced under Earth's surface is called

____________________.

Match the correct description with the correct term. Write the letter in the space provided.

_____ **6.** place where two tectonic plates move away from each other

_____ **7.** vertical fractures that occur when tectonic plates pull apart

_____ **8.** place where two tectonic plates collide

_____ **9.** the movement of one tectonic plate under another tectonic plate

_____ **10.** mountain chain created by underwater lava

- **a.** mid-ocean ridge
- **b.** divergent boundary
- **c.** convergent boundary
- **d.** subduction
- **e.** fissures

11. What are hot spots?

____

____

12. What are mantle plumes?

____

____

HOW MAGMA FORMS

13. Magma forms in the deeper regions of Earth's ________________ and the upper regions of Earth's ________________.

14. Changes in ________________ and ________________ cause magma to form.

15. What happens to rock when Earth's temperature increases?

____

____

16. Why is the expansion of rock important for magma to form?

____

____

17. Explain how the addition of fluids to hot rock causes magma to form.

____

____

Name ______________________ Class ______________ Date ____________

Skills Worksheet

Directed Reading B

Section: Types of Volcanoes (pp. 270–277)

1. The formation and composition of magma determines how different types of

______________________ are created.

VOLCANOES AT DIVERGENT BOUNDARIES

______ **2.** Which of the following statements is true?
- **a.** Lava that is low in silica has a thin, runny consistency.
- **b.** Lava that is low in silica traps water and gas bubbles.
- **c.** Lava that is low in silica causes explosive eruptions.
- **d.** Lava that is low in silica causes pyroclastic flows.

______ **3.** Mid-ocean ridges are
- **a.** groups of tidal waves.
- **b.** formed when two tectonic plates collide.
- **c.** always located near hot springs.
- **d.** underwater volcanic mountain chains.

4. An area of deep cracks that forms when two tectonic plates are pulling away

from each other is called a(n) ______________________.

5. Magma that flows onto Earth's surface is called ______________________.

6. What is the term *mafic* used to describe?

____

____

7. Why is Iceland getting larger?

____

____

VOLCANOES AT HOT SPOTS

______ **8.** Lava at hot spots originates
- **a.** in the oceanic lithosphere.
- **b.** in Earth's mantle.
- **c.** from tidal waves.
- **d.** in Earth's atmosphere.

______ **9.** How is lava released through a volcano?
- **a.** through the magma chamber
- **b.** through vents in Earth's surface
- **c.** through Earth's core
- **d.** through mid-ocean ridges

10. A hot spot forms when a(n) ______________________ moves over a(n) ______________________.

11. Shield volcanoes usually form at ______________________.

12. How do shield volcanoes form?

____

____

13. Where are magma chambers located?

____

VOLCANOES AT CONVERGENT BOUNDARIES

14. When water lowers the melting temperature of rock, the rock ______________________.

15. What is the term *felsic* used to describe?

____

____

16. Why is silica-rich magma associated with explosive eruptions?

____

____

17. What is pyroclastic material?

____

____

18. When large amounts of hot ash, dust, and gases are ejected from a volcano, the result is a dangerous type of flow called a(n) ______________________.

19. Pyroclastic materials can race downhill at speeds of more than ______________________.

Match the correct description with the correct term. Write the letter in the space provided.

_____ **20.** volcanoes formed from layers of lava and pyroclastic materials

_____ **21.** another name for composite volcanoes

_____ **22.** volcanoes made entirely of pyroclastic material

a. composite volcanoes

b. cinder cone volcanoes

c. stratovolcanoes

Name ______________________ Class ______________ Date ______________

Skills Worksheet

Directed Reading B

Section: Effects of Volcanic Eruptions (pp. 278–281)

1. A large ______________________ explosion can affect the global climate.

NEGATIVE IMPACTS OF VOLCANIC ERUPTIONS

______ 2. How does a volcanic eruption affect climate change?
- **a.** Burned land creates dry conditions.
- **b.** The sun shines brighter, causing temperatures to rise.
- **c.** Ash blocks sunlight, causing temperatures to drop.
- **d.** Volcanic eruptions don't cause climate changes.

______ 3. What are two global impacts of a temperature change caused by a volcanic eruption?
- **a.** crop failure and starvation
- **b.** farm equipment failure and less drinking water
- **c.** stronger building materials and cleaner water
- **d.** fuel-efficient cars and stronger crops

4. The blast from a(n) ______________________ eruption is the most destructive type of eruption.

5. What is a lahar?

____

____

6. What is a specific health problem that volcanic ash can cause?

____

7. How can large volcanic eruptions cause global temperatures to decrease?

____

____

BENEFITS OF VOLCANIC ERUPTIONS

______ 8. One benefit that volcanoes provide is
- **a.** fresh drinking water.
- **b.** construction materials.
- **c.** new medicines.
- **d.** warmer temperatures.

______ **9.** The liquid in volcanic rocks that is heated by magma is called
a. volcanic flow.
b. geothermal water.
c. acid rain.
d. drinking water.

______ **10.** Geothermal energy is used
a. to keep roads free of ice.
b. to eliminate traffic jams.
c. to create habitats for animals.
d. to bury crops.

______ **11.** Where are the world's greatest consumers of geothermal energy located?
a. Tokyo, Japan
b. Juneau, Alaska
c. Madrid, Spain
d. Reykjavik, Iceland

______ **12.** What volcanic rocks are commonly used in the construction of roads and bridges?
a. ash and lava
b. silica sand and ash
c. basalt and pumice
d. basalt and ash

13. Why is volcanic soil a benefit of volcanic eruptions?

____

____

14. Why is pumice added to soil?

____

____

Name ______________________ Class ______________ Date ______________

Skills Worksheet

Vocabulary and Section Summary B

Why Volcanoes Form

VOCABULARY

After you finish reading the section, try this puzzle! Look at the clues below, and write the terms being described in the blanks below. Then, copy the numbered letters into the corresponding blanks at the bottom of the page to reveal the answer to the bonus question.

1. area where two tectonic plates pull apart

2. a vent in Earth's surface through which magma and gases pass

3. a set of deep fractures

4. volcanically active places that are not located near tectonic plate boundaries

5. area where two tectonic plates collide

6. the process in which one tectonic plate moves beneath another plate

7. liquid rock produced under Earth's surface

1. ___ ___ ___ ___ ___ ___ ___ ___ ___
(8 under the 2nd blank)

___ ___ ___ ___ ___ ___ ___ ___
(9 under the 7th blank)

2. ___ ___ ___ ___ ___ ___ ___
(3 under the 6th blank)

3. ___ ___ ___ ___ ___ ___ ___ ___
(7 under the 1st blank)

4. ___ ___ ___   ___ ___ ___ ___ ___
(5 under the 6th blank)

5. ___ ___ ___ ___ ___ ___ ___ ___ ___ ___
(10 under the 5th blank)

___ ___ ___ ___ ___ ___ ___ ___
(1 under the 7th blank)

6. ___ ___ ___ ___ ___ ___ ___ ___ ___ ___
(2 under the 8th blank)

7. ___ ___ ___ ___ ___
(4 under the 3rd blank)

What are the plate boundaries surrounding the Pacific Ocean called?

___ ___ ___ ___   ___ F   ___ ___ ___ ___
1 2 3 4   5 6   7 8 9 10

SECTION SUMMARY

Read the following section summary.

- A volcano is a vent or fissure in Earth's surface through which magma and gases pass.
- Most volcanoes are located at tectonic plate boundaries.
- Volcanic activity occurs at divergent plate boundaries, convergent plate boundaries, and hot spots.
- Magma forms when the temperature of a rock increases, when the pressure on a rock decreases, or when water lowers the melting temperature of a rock.

Name ______________________ Class ______________ Date ______________

Skills Worksheet

Vocabulary and Section Summary B

Types of Volcanoes

VOCABULARY

After you finish reading the section, try this puzzle! Use the clues below to fill in the missing letters in the volcano puzzle below. All the words are spelled horizontally.

1. magma that flows onto Earth's surface
2. term used to describe magma that is rich in magnesium and iron
3. term used to describe magma that is rich in feldspars and silica
4. a set of deep cracks that forms at a divergent boundary
5. pipelike structures within a volcano through which magma travels
6. area deep underground where molten rock collects
7. another name for composite volcanoes
8. volcanoes that form at hot spots from many nonexplosive eruptions
9. process in which new sea floor forms as older sea floor is pulled apart

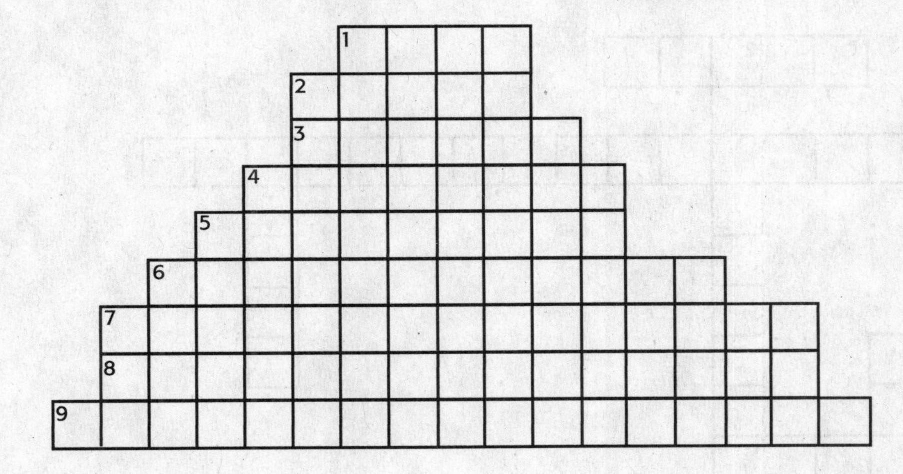

SECTION SUMMARY

Read the following section summary.

- Mafic lava erupts quietly through cracks, or fissures, in the lithosphere at divergent boundaries.
- At hot spots, continuous eruptions of mafic magma form chains of volcanoes above mantle plumes.
- Shield volcanoes form from the mafic lava erupted at hot spots.
- At convergent boundaries, eruptions of silica-rich magma are often explosive.
- Composite volcanoes form from the felsic lava erupted at convergent boundaries.

Name ______________________ Class ______________ Date ____________

Skills Worksheet

Vocabulary and Section Summary B

Effects of Volcanic Eruptions

VOCABULARY

After you finish reading the section, try this puzzle! Use the clues below to solve the crossword puzzle.

ACROSS

1. liquid rock produced under Earth's surface

3. large amounts of hot ash, dust, and toxic gases ejected from a powerful volcanic eruption

6. magma that flows onto Earth's surface

DOWN

2. the liquid found in volcanic rocks that is heated by magma

4. a mudflow that forms when volcanic ash and debris mix with water during a volcanic eruption

5. a vent or fissure in Earth's surface through which magma and gases are expelled

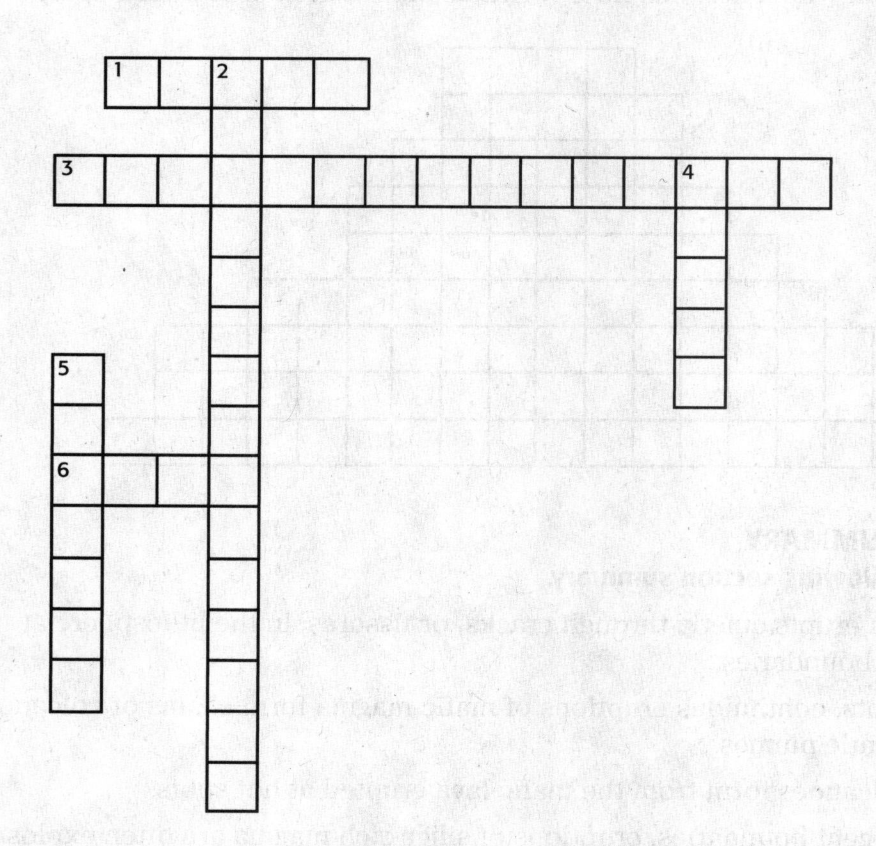

SECTION SUMMARY

Read the following section summary.

- Volcanic eruptions can have local effects on humans and on wildlife habitats.
- When ash and gases from a large volcanic eruption spread around the planet, they may absorb and scatter enough sunlight to cause a temporary decrease in the average global temperature.
- Benefits that volcanoes provide to humans and to the environment include fertile soil, a renewable energy source, and construction materials.

Name ______________________ Class ______________ Date ____________

Skills Worksheet

Directed Reading B

Section: Weathering (pp. 298–303)

1. What is the process of weathering?

____

____

MECHANICAL WEATHERING

_____ **2.** Mechanical weathering is the breakdown of rock into smaller pieces by
- **a.** warm weather.
- **b.** cold weather.
- **c.** chemical processes.
- **d.** physical means.

_____ **3.** Ice, wind, water, gravity, plants, and animals can all be agents of
- **a.** mechanical weathering.
- **b.** chemical weathering.
- **c.** chemical processes.
- **d.** abrasion.

_____ **4.** The alternate freezing and thawing of soil and rock is called
- **a.** frost action.
- **b.** abrasion.
- **c.** oxidation.
- **d.** contraction.

_____ **5.** Ice wedging occurs when water filling a crack in a rock
- **a.** thaws and flows out.
- **b.** freezes and contracts.
- **c.** freezes and expands.
- **d.** thaws and causes abrasion.

6. The grinding and wearing away of rock surfaces through the mechanical action of other rocks or sand particles is called ______________________.

7. Rocks that have been shaped by blowing sand are called

______________________.

Match the correct description with the correct term. Write the letter in the space provided.

_____ **8.** one rock falling against another rock

_____ **9.** pebbles bumping against each other in a stream

_____ **10.** sand blowing against rock

a. wind
b. gravity
c. water

11. What is the relationship between the erosion of surface material of rock and the process of exfoliation?

____

____

____

12. How can a plant break a rock?

____

____

____

13. In what way can an animal cause mechanical weathering?

____

____

____

CHEMICAL WEATHERING

_____ **14.** The process by which rocks break down as a result of chemical reactions is called
a. abrasion.
b. mechanical weathering.
c. chemical weathering.
d. acid precipitation.

_____ **15.** Rocks can break down when chemicals within the rock dissolve due to the action of
a. abrasion.
b. gravity.
c. water.
d. wind.

______ **16.** Rain, sleet, or snow that contains a high concentration of acids is called
a. mechanical weathering.
b. acid precipitation.
c. chemical weathering.
d. mechanical abrasion.

______ **17.** Compounds formed by the combination of burning fossil fuels and water in the atmosphere are
a. phosphoric acids.
b. acetic acids.
c. carbon monoxide gases.
d. acids.

______ **18.** Acid precipitation can result from
a. mineral bonding.
b. groundwater weathering.
c. fossil fuels burning.
d. mechanical weathering.

19. Over a long period of time, acids in the groundwater can cause chemical weathering of limestone. This weathering can form a(n) ______________________.

20. How do living things cause chemical weathering?

____

____

____

21. The chemical reaction in which an element, such as iron, combines with oxygen to form an oxide is called ______________________.

22. When oxygen in the air reacts with metal, oxidation occurs and causes the metal to ______________________.

Name ______________________ Class ______________ Date ____________

Skills Worksheet

Directed Reading B

Section: Rates of Weathering (pp. 304–307)

1. What three factors determine the rate at which rock weathers?

____

____

____

DIFFERENTIAL WEATHERING

______ **2.** The process by which softer, less weather resistant rocks wear away at a faster rate than harder, more weather resistant rocks is called
a. mechanical weathering.
b. chemical weathering.
c. differential weathering.
d. acid precipitation.

______ **3.** Scientists believe Devils Tower once was a mass of
a. molten rock.
b. granite boulders.
c. underground caves.
d. hard mud.

______ **4.** The landform called Devils Tower is made up of the hard, weather resistant rocks that remained after the softer rocks
a. became lava.
b. oxidized.
c. slid downhill.
d. wore away.

THE SURFACE AREA OF ROCKS

______ **5.** The part of a rock where weathering takes place is
a. throughout the rock.
b. on its outer surface.
c. between the core and surface.
d. inside the core.

______ **6.** Because a large rock has a large volume, it will
a. weather unevenly.
b. weather quickly.
c. not weather at all.
d. weather slowly.

7. Small rocks weather more quickly than large rocks because they have more surface area compared to their ______________________.

WEATHERING AND CLIMATE

8. The average weather condition in an area over a long period of time is called ______________________.

9. At ______________________ temperatures, many chemical reactions occur more rapidly.

10. Chemical reactions such as oxidation occur more quickly in a climate that is ______________________ and ______________________.

OTHER FACTORS THAT AFFECT WEATHERING

______ **11.** Weathering occurs faster at high elevations because

- **a.** wind increases but precipitation does not.
- **b.** precipitation increases but wind does not.
- **c.** wind and precipitation decrease.
- **d.** wind and precipitation increase.

12. How does gravity contribute to the weathering of steep mountain slopes?

____

____

13. What are some actions of living things that contribute to chemical and mechanical weathering?

____

____

Name ______________________________ Class ________________ Date ______________

Skills Worksheet

Directed Reading B

Section: From Bedrock to Soil (pp. 308–313)

THE SOURCE OF SOIL

Match the correct description with the correct term. Write the letter in the space provided.

______ **1.** soil that is carried away from its parent rock by wind, water, ice, or gravity

______ **2.** the layer of rock beneath the soil

______ **3.** a loose mixture of small mineral fragments, organic material, water, and air that can support the growth of vegetation

______ **4.** the rock that breaks down into mineral fragments that form a soil

______ **5.** soil that remains above its parent rock

a. soil
b. transported soil
c. residual soil
d. parent rock
e. bedrock

SOIL PROPERTIES

Match the correct description with the correct term. Write the letter in the space provided.

______ **6.** kinds and amounts of materials in the soil

______ **7.** soil quality that is based on the proportions of soil particles

______ **8.** water's ability to move through the soil

______ **9.** soil's ability to hold nutrients and to supply nutrients to a plant

a. soil texture
b. soil composition
c. soil fertility
d. infiltration

10. The organic material formed in soil from the decayed remains of plants and animals is called ____________________.

11. Because of the way soil forms, soil often ends up in a series of horizontal layers called ____________________.

12. The top layer of soil, commonly called ____________________, usually contains more humus than the layers below it.

13. What is the pH scale used to measure?

____

14. What influences how nutrients dissolve in the soil?

____

SOIL AND CLIMATE

Match the correct description with the correct term. Write the letter in the space provided.

_____ **15.** climate with heavy rain that removes precious nutrients from the topsoil

_____ **16.** climate with low temperatures and slow soil formation, resulting in thin soil and little plant and animal life

_____ **17.** climate with little rainfall and a low ability to support plant and animal life

_____ **18.** climate with enough rain to cause a high level of chemical weathering, but not so much that the nutrients are removed from the soil

a. desert

b. temperate forest and grassland

c. tropical rain forest

d. arctic areas

19. Lush plant growth takes large amounts of nutrients from the soil in which climate?

____

20. When dead plants decay quickly in warm soil, they produce rich

_______________.

21. Why are tropical soils generally poor and thin?

____

22. High levels of salt are sometimes found in the soil in which climate?

____

23. A combination of low temperature and little precipitation in a(n)

_______________ climate often means a lack of plants and animals and a limited amount of nutrients that become part of the soil.

24. Which climate has soil that is the most productive in the world and can support many organisms?

____

Name ______________________ Class ______________ Date ____________

Skills Worksheet

Directed Reading B

Section: Soil Conservation (pp. 314–317)

1. A resource like soil that can take thousands of years to replace is considered a(n) ______________________.

2. Maintaining the fertility of soil by preventing erosion and nutrient loss is called ______________________.

THE IMPORTANCE OF SOIL

3. Soil provides nutrients and moisture to ______________________, which grow directly in it.

4. Even animals that eat plants benefit when ______________________ is rich in nutrients.

5. The region where a plant or animal lives is called its ______________________.

6. Soil's ability to ______________________ allows plants to get the moisture they need.

SOIL DAMAGE AND LOSS

7. Soil can be damaged and lose its ______________________ from overuse or overgrazing.

8. Without plants to help hold and cycle water, an infertile area can become like a desert, a process called ______________________.

9. The process by which wind, water, ice, or gravity transports soil and sediment from one location to another is called ______________________.

SOIL CONSERVATION ON FARMLAND

10. Name three methods by which farmers can help prevent erosion.

____

____

____

11. Crops that are planted between harvests to replace certain nutrients and prevent erosion are called ______________________.

12. The practice of planting different crops each year to slow down nutrient depletion and support more plants is called ______________________.

13. How does crop rotation slow down nutrient depletion?

____

____

____

Name ______________________ Class ______________ Date ____________

Skills Worksheet

Vocabulary and Section Summary B

Weathering

VOCABULARY

After you finish reading the section, try this puzzle! Use the clues below to solve the crossword puzzle.

ACROSS

5. rain, sleet, or snow that contains a high concentration of acids

6. the process by which rocks break down as a result of chemical reactions

DOWN

1. the process by which rock surfaces break down into smaller pieces by physical means

2. the grinding and wearing away of rock surfaces through the mechanical action of other rock or sand particles

3. the process by which sheets of rock peel away when pressure is removed

4. the process by which atmospheric and environmental agents decompose rocks

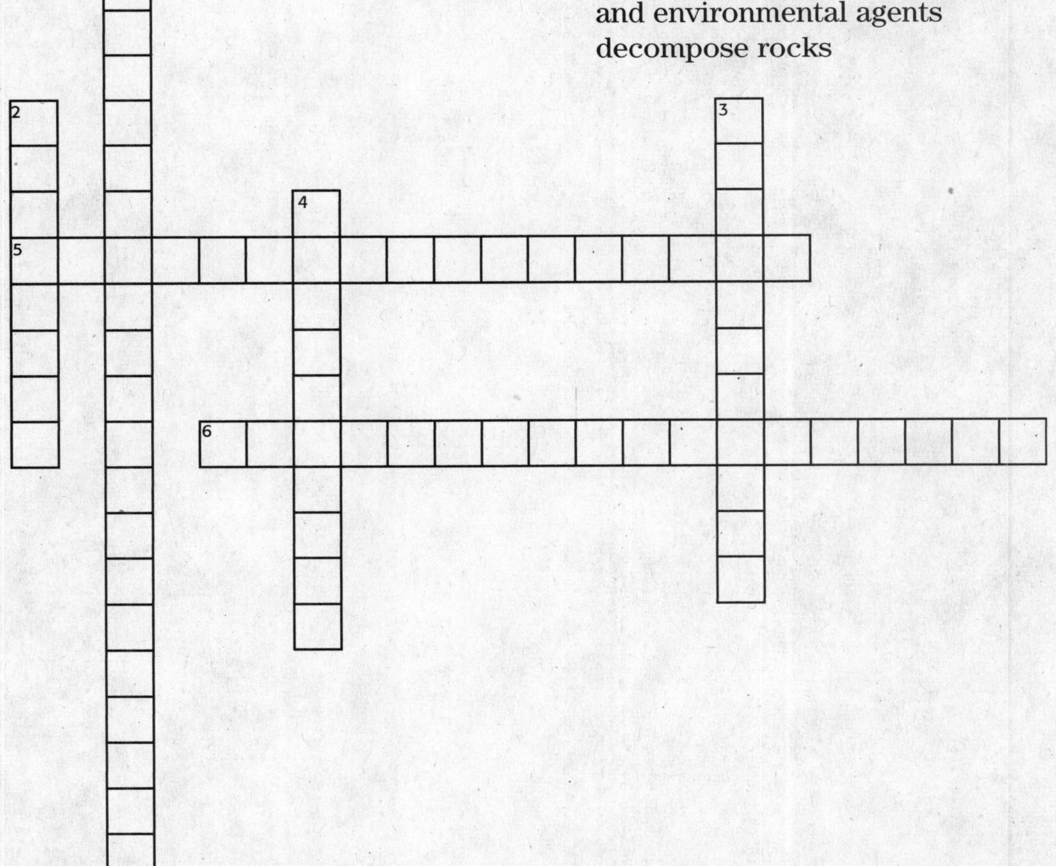

SECTION SUMMARY

Read the following section summary.

- Ice wedging is a form of mechanical weathering in which water seeps into cracks in rock and then freezes and expands.
- Wind, water, and gravity cause mechanical weathering by abrasion.
- Animals and plants cause mechanical weathering by mixing the soil and breaking apart rocks.
- Water, acids, and oxygen in the air chemically weather rock by reacting chemically with elements in the rock.
- Oxidation is the process by which oxygen from the air reacts with iron in rocks.

Name ______________________ Class ______________ Date ____________

Skills Worksheet

Vocabulary and Section Summary B

Rates of Weathering

VOCABULARY

After you finish reading the section, try this puzzle! In the space provided, write the term described. Then, find the words in the word search puzzle on the next page. Words are hidden vertically, horizontally, diagonally, and backward.

______________________ **1.** the process by which softer, less weather resistant rocks wear away at a faster rate than harder, more weather resistant rocks do

______________________ **2.** the natural process by which atmospheric and environmental agents, such as wind, rain, and temperature changes, disintegrate and decompose rocks

______________________ **3.** the breakdown of rock into smaller pieces by physical means

______________________ **4.** the grinding and wearing away of rock surfaces through the mechanical action of other rock or sand particles

______________________ **5.** the process by which rocks break down as a result of chemical reactions

T	Y	B	D	F	A	M	S	T	D	I	W	M	N	U	I	F	D	C	G
E	F	F	I	H	R	Q	L	K	D	W	F	D	F	I	G	B	P	N	C
R	W	H	F	C	Y	E	D	O	W	U	Q	G	E	U	U	M	I	X	Y
O	H	L	F	W	Z	X	Z	D	X	E	E	U	J	X	A	R	H	Y	A
P	T	V	E	P	H	J	C	W	A	T	U	Z	C	L	E	L	Z	M	M
M	J	F	R	I	K	S	Z	P	B	B	N	J	M	H	U	N	K	S	F
P	K	U	E	A	Y	K	Y	I	X	E	R	Q	T	C	E	W	O	X	M
G	B	U	N	K	L	I	S	Q	Y	O	D	A	N	I	O	U	T	H	Z
S	B	K	T	G	G	Y	H	V	U	U	E	R	S	G	D	W	G	N	P
Y	G	N	I	R	E	H	T	A	E	W	L	A	C	I	M	E	H	C	G
Z	L	I	A	A	J	Q	A	M	L	P	M	C	R	Y	O	T	O	P	L
J	W	W	L	C	Q	P	L	A	M	S	C	E	W	B	Z	N	B	Z	P
F	R	N	W	I	H	A	C	B	G	H	Y	R	K	X	Y	G	J	I	D
S	Q	Y	E	U	Z	I	L	W	D	T	D	X	X	C	I	S	J	I	A
J	X	D	A	H	N	K	U	V	R	H	J	Q	V	J	P	R	L	G	G
U	K	Y	T	A	V	N	N	R	H	F	K	L	H	S	N	J	W	I	D
L	R	G	H	B	V	A	G	K	C	O	Q	P	R	Q	K	M	P	D	Z
V	J	C	E	E	C	C	U	T	A	H	D	Y	D	H	I	Y	K	R	Y
G	E	G	R	K	U	M	Q	H	C	X	L	R	P	O	H	E	Q	E	K
M	H	V	I	I	G	H	F	E	C	C	B	B	H	Z	H	T	C	F	A
B	G	J	N	S	S	X	E	K	J	W	Q	A	U	Z	G	Q	B	F	X
P	L	B	G	H	I	H	T	V	G	H	P	B	K	M	F	V	Q	K	X

SECTION SUMMARY

Read the following section summary.

- Hard rocks weather more slowly than soft rocks.
- The larger the surface area–to-volume ratio of a rock is, the faster the rock will wear down.
- Chemical weathering occurs faster in warm, wet climates than in hot, dry climates.
- Rates of weathering are affected by elevation, by the slope of the ground, and by living things.

Name ______________________ Class ______________ Date ____________

Skills Worksheet

Vocabulary and Section Summary B

From Bedrock to Soil

VOCABULARY

After you finish reading the section, try this puzzle! Unscramble the letters at the end of each description to find the word that is being described.

1. a loose mixture of rock fragments, organic material, water, and air that can support the growth of vegetation: LSIO

____

2. a rock formation that is the source of soil: NRAPET RKCO

____

3. the layer of rock beneath soil: KODCBER

____

4. dark, organic material formed in soil from the decayed remains of plants and animals: USMHU

____

5. soil that remains above its parent rock: DSEILARU SLOI

____

6. soil that is carried away from its parent rock by wind, water, ice, or gravity: TSRNETPAORD LOSI

____

7. the relative amounts of soil particles of different sizes: XTEREUT ISLO

____

8. the ability of water to move into the soil: OAINTLTINFRI

____

9. the ability of soil to hold nutrients and to supply nutrients to plants: OSIL ITIETLFRY

____

10. a series of horizontal layers in which soil forms, with humus-rich soil generally on top: RNROZIHOS

____

11. the top layer of soil: IOSOLPT

____

12. the removal of substances that can be dissolved from rock, ore, or layers of soil due to the passing of water: EHLAC

____

SECTION SUMMARY

Read the following section summary.

- Soil forms from the weathering of bedrock.
- Soil texture affects how soil can be worked for farming and how well water passes through soil.
- The ability of soil to provide nutrients so that plants can survive and grow is called *soil fertility.*
- The pH of a soil influences which nutrients plants can take up from the soil.
- Different climates have different types of soil, depending on the temperature and rainfall.
- The characteristics of soil affect the number and types of organisms that an area can support.

Name ______________________ Class ______________ Date ____________

Skills Worksheet

Vocabulary and Section Summary B

Soil Conservation

VOCABULARY

After you finish reading the section, try this puzzle! The underlined words are missing all their vowels. Write the completed words in the spaces provided.

1. Plowing across the slope of a hill, called CNTR PLWNG, slows the downhill flow of water and reduces erosion.

   ____

2. Due to the process of DSRTFCTN, an area becomes a desert because it lacks the plants needed to hold and help cycle water.

   ____

3. SL CNSRVTN is a method used to maintain the fertility of the soil by protecting the soil from erosion and nutrient loss.

   ____

4. Farmers practice CRP RTATN by planting different crops in different years in the same field to slow nutrient depletion.

   ____

5. TRRCNG is the practice of changing one steep field into a series of small, flat fields that prevent the rapid runoff of water.

   ____

6. The process by which wind, water, ice, or gravity transports soil and sediment from one location to another is called RSN.

   ____

7. In the practice of N-TLL FRMNG, old stalks are left to provide cover from rain, thus reducing water runoff and slowing soil erosion.

   ____

8. CVR CRPS are planted between harvests to allow the soil to replace certain nutrients and to prevent erosion.

   ____

SECTION SUMMARY

Read the following section summary.

- Soil forms slowly over hundreds or thousands of years. Therefore, soil is considered a nonrenewable resource.
- Soil is important because plants grow in soil, animals live in soil, and water is stored in soil.
- Soil can be eroded by water running downhill or by wind.
- Soil erosion and soil damage can be prevented by no-till farming, contour plowing, terracing, using cover crops, and practicing crop rotation.

Name ______________________ Class ______________ Date ______________

Skills Worksheet

Directed Reading B

Section: Shoreline Erosion and Deposition (pp. 332–339)

______ **1.** When waves crash into rock over a long period of time,
- **a.** the waves gain energy.
- **b.** the rock becomes sand.
- **c.** the waves lose energy.
- **d.** the rock moves seaward.

______ **2.** The boundary between land and water is a(n)
- **a.** sea stack.
- **b.** estuary.
- **c.** wave.
- **d.** shoreline.

WAVE ENERGY

3. The strength of the wind and the length of time the wind blows affect the size of a(n) ______________________.

______ **4.** Waves travel in groups called
- **a.** wave trains.
- **b.** ocean waves.
- **c.** wave periods.
- **d.** surf.

______ **5.** Breaking waves are called
- **a.** wave trains.
- **b.** ocean waves.
- **c.** wave periods.
- **d.** surf.

6. The length of time between breaking waves is the ______________________.

7. Explain how waves can turn rock into sand.

____

____

____

____

____

WAVE EROSION

Match the correct description with the correct term. Write the letter in the space provided.

______	**8.** These are offshore columns of rock.	**a.** headlands
______	**9.** These are formed when cliffs of hard rock erode more slowly than the surrounding softer rock does.	**b.** wave-cut terraces
______	**10.** These are formed when waves cut holes into weak rock along the base of sea cliffs.	**c.** sea stacks
______	**11.** These are formed when a sea cliff wears back.	**d.** sea arches
______	**12.** These are formed by waves eroding a sea cave, cutting completely through the rock.	**e.** sea caves

WAVE DEPOSITS

______ **13.** An area of shoreline made up of material deposited by waves is called a(n)

a. headland. **c.** wave-cut terrace.
b. beach. **d.** island.

14. Why aren't all beaches made up of light-colored sand?

____

____

____

15. A water current that carries pieces of sand and rock away from the shore is

a(n) ______________________.

16. A water current that moves sand in a zigzag pattern along a beach

is a(n) ______________________.

17. An underwater ridge of sand, gravel, or shell is

a(n) ______________________.

18. An exposed sandbar connected to the shoreline is called

a(n) ______________________.

19. Why are some of California's offshore islands home to rare species?

____

____

Name ______________________ Class ______________ Date ____________

Skills Worksheet

Directed Reading B

Section: Wind Erosion and Deposition (pp. 340–343)

WIND EROSION

______ **1.** How do plants reduce wind erosion?
a. Plants shade the soil.
b. Plant roots hold sand and soil in place.
c. Plants break down and make soil.
d. Plant roots help break up soil.

Match the correct description with the correct term. Write the letter in the space provided.

______ **2.** wind causing sand-sized particles to skip and bounce

______ **3.** wind erosion in which fine, dry soil particles are blown away

______ **4.** grinding and wearing down of rock by other rock or sand

a. abrasion
b. saltation
c. deflation

WIND-DEPOSITED MATERIALS

______ **5.** Wind carries particles called
a. water.
b. rocks.
c. sediment.
d. desert pavement.

______ **6.** A mound of wind-deposited sand that moves as a result of the wind is called a(n)
a. dune.
b. boulder.
c. deflation hollow.
d. desert pavement.

Place the five steps involved in forming a sand dune in order from first to last. Write the appropriate number in the space provided.

______ **7.** Material collects and creates an additional obstacle.

______ **8.** Wind deposits more material, forming a mound or dune.

______ **9.** Slowing wind drops the heavier particles.

______ **10.** Wind hits a rock, plant, or other object and slows down.

______ **11.** The original object eventually becomes buried.

12. Name the largest desert sand dune system in California.

____

13. Explain how dunes migrate in the direction of the wind.

____

____

____

Name ______________________ Class ______________ Date ______________

Skills Worksheet

Directed Reading B

Section: Erosion and Deposition by Ice (pp. 344–347)

1. What is a glacier?

____

____

2. What type of glacier forms in mountainous areas?

____

____

3. What is a continental glacier?

____

____

GLACIERS—RIVERS OF ICE

Place the four steps involved in formation of glaciers in order from first to last. Write the appropriate number in the space provided.

______ **4.** Gravity pulls on the compressed ice, setting it in motion.

______ **5.** Snow builds up year after year in polar regions and high elevations.

______ **6.** The layers of packed snow become ice.

______ **7.** The weight of the top layers of snow compresses and packs the lower layers.

8. Why do glaciers flow like "rivers of ice"?

____

____

9. How and where are valley glaciers formed?

____

____

____

10. Name the two ways glaciers move.

____

____

LANDFORMS CREATED BY GLACIERS

Match the correct description with the correct term. Write the letter in the space provided.

______ **11.** smaller glacial valleys that join the deeper main valley

______ **12.** bowl-like depressions where glacial ice cuts into the mountain walls

______ **13.** pyramid-shaped peaks that form when three or more cirque glaciers erode a mountain

______ **14.** jagged ridges that form between two or more cirques cutting into the same mountain

______ **15.** valleys formed when a glacier erodes a river valley from the valley's original V shape

a. horns
b. hanging valleys
c. arêtes
d. cirques
e. U-shaped valleys

TYPES OF GLACIAL DEPOSITS

16. All material carried and deposited by glaciers is called

______________________.

17. What are the two main types of glacial drift?

____

____

18. Unsorted rock material that is deposited directly by the ice when it melts is called ______________________.

19. A glacial deposit that is sorted and layered by streams or meltwater is called ______________________.

Identify each of the following descriptions as stratified drift (S) or till deposit (T), and write *S* or *T* in the space provided.

______ **20.** Sediment built around a block of ice eventually forms a kettle.

______ **21.** Ridges form along the edges of glaciers.

______ **22.** Streams from glaciers carry sorted material to an area called an outwash plain.

______ **23.** Moraines form ridges along the edges of glaciers.

Match the correct definition with the correct term. Write the letter in the space provided.

______ **24.** forms from till left beneath a glacier

______ **25.** forms when sediment is dropped at the front of a glacier

______ **26.** forms along the sides of a glacier

______ **27.** forms when valley glaciers that have moraines along each side meet

a. medial moraine

b. lateral moraine

c. ground moraine

d. terminal moraine

Name ______________________ Class ______________ Date ____________

Skills Worksheet

Directed Reading B

Section: Erosion and Deposition by Mass Movement (pp. 348–351)

1. What causes ice, rocks, and soil to move down a slope?

____

2. What is mass movement?

____

____

____

____

ANGLE OF REPOSE

3. What is the angle of repose?

____

____

____

____

4. What are three characteristics of a material that will affect its angle of repose?

____

____

____

RAPID MASS MOVEMENT

Match the correct definition with the correct term. Write the letter in the space provided.

_____ **5.** loose rocks falling down a steep slope

_____ **6.** sudden, rapid movement of rock and soil down a slope

_____ **7.** slow downhill movement of weathered rock

_____ **8.** rapid movement of a large mass of mud

a. landslide

b. creep

c. mudflow

d. rock fall

9. The most common type of landslide is called a(n)

______________________.

10. What are three factors that affect creep?

____

____

____

MASS MOVEMENT AND LAND USE

11. Why should the likelihood of mass movement be considered in a land-use decision?

____

____

____

Name ______________________ Class ______________ Date ____________

Skills Worksheet

Vocabulary and Section Summary B

Shoreline Erosion and Deposition

VOCABULARY

After you finish reading the section, try this puzzle! The underlined terms below are missing all their vowels. Use the clues below to write the corrected words in the spaces provided.

1. The place where land and a body of water meet is called a(n) SHRLN.

 ____

2. A(n) BCH is any area of shoreline that is made up of materials deposited by waves.

 ____

3. A subsurface current near shore that can pull objects out to sea is a(n) NDRTW.

 ____

4. A(n) LNGSHR CRRNT travels near and parallel to the shoreline.

 ____

5. Waves move in groups called WV TRNS.

 ____

6. A(n) WV PRD is the time between waves in a wave train.

 ____

7. Breaking waves are known as SRF.

 ____

8. When waves erode and undercut rock to make steep slopes, S CLFFS are formed.

 ____

9. A(n) SNDBR is an underwater or exposed ridge of sand, gravel, or shell material.

 ____

SECTION SUMMARY

Read the following section summary.

- A wave is a disturbance in the water that can be caused by wind.
- As waves break against a shoreline, their energy breaks rocks down into sand.
- Six shoreline features that are created by wave erosion are sea cliffs, sea stacks, sea caves, sea arches, headlands, and wave-cut terraces.
- Beaches are made from material deposited by rivers and waves.
- California beaches can be rocky or sandy and can have different mineral compositions.
- Longshore currents cause sand to move in a zigzag pattern along the shore.
- Longshore currents can deposit eroded sediment offshore.

Name ______________________ Class ______________ Date ______________

Skills Worksheet

Vocabulary and Section Summary B

Wind Erosion and Deposition

VOCABULARY

After you finish reading the section, try this puzzle! Use the clues below to solve the crossword puzzle.

ACROSS

2. a mound of wind-deposited sand

3. the grinding and wearing away of rock surfaces by other rock or sand particles

4. a type of wind erosion in which fine, dry particles are blown away

DOWN

1. movement of sand by short jumps and bounces that is caused by wind

2. a surface made of pebbles and small broken rocks

SECTION SUMMARY

Read the following section summary.

- Areas that have little plant cover and desert areas that are covered with fine rock material are more vulnerable to wind erosion than other areas are.
- Saltation is the process in which sand-sized particles move in the direction of the wind.
- Desert pavement, deflation hollows, and dunes are landforms that are created by wind erosion and deposition.
- Dunes move in the direction that the wind blows.

Name ______________________ Class ______________ Date __________

Skills Worksheet

Vocabulary and Section Summary B

Erosion and Deposition by Ice

VOCABULARY

After you finish reading the section, try this puzzle! Fill in the blanks with the correct terms. Then, use the words to complete the word search puzzle on the next page. Words may appear horizontally, vertically, backward, or diagonally.

1. A(n) ______________________ is a large mass of moving ice.

2. The general term used to describe all material carried and deposited by glaciers is ______________________ ______________________.

3. When a glacier melts, the unsorted rock material that is deposited directly by the ice is called______________________.

4. The most common type of till deposit is a(n) ______________________.

5. A glacial deposit that is sorted into layers based on the size of the rock material is called ______________________ ______________________.

6. A(n) ______________________ ______________________ is the broad area in front of a glacier, where streams deposit sorted material.

7. A depression that commonly fills with water to form a lake or a pond is called a(n) ______________________.

S	W	E	O	Y	T	A	G	M	H	E	O	G	O	Q
I	T	W	H	E	H	K	R	I	K	U	Q	L	N	G
H	G	R	D	H	S	N	Z	C	T	P	G	A	X	E
G	B	N	A	R	Y	P	X	W	N	S	Q	C	I	V
E	D	Q	R	T	G	Y	A	D	M	A	H	I	H	V
L	Y	Q	H	G	I	S	T	P	Q	P	O	A	M	D
T	V	Y	P	R	H	F	L	G	S	A	S	L	T	P
T	V	P	R	P	E	K	I	A	L	I	F	D	H	Q
E	S	G	L	U	N	I	S	E	N	I	A	R	O	M
K	G	A	I	Y	F	Z	C	H	D	Z	S	I	Z	Y
E	I	P	Z	E	X	N	I	A	D	D	J	F	G	Y
N	Z	B	X	Y	M	H	H	C	L	L	R	T	K	S
S	U	N	R	W	G	M	G	E	L	G	S	I	K	U
Y	I	A	U	K	U	E	U	I	L	U	C	S	F	Y
L	M	H	B	F	N	Z	T	E	R	A	V	P	S	T

SECTION SUMMARY

Read the following section summary.

- Glaciers are masses of moving ice that shape the landscape by eroding and depositing material.
- Glaciers move by sliding or by flowing.
- Alpine glaciers can carve cirques, arêtes, horns, U-shaped valleys, and hanging valleys.
- Two types of glacial deposits are till and stratified drift.
- Deposition of sediment by glaciers can form several landforms, including kettles.

Name ______________________ Class ______________ Date ______________

Skills Worksheet

Vocabulary and Section Summary B

Erosion and Deposition by Mass Movement

VOCABULARY

After you finish reading the section, try this puzzle! Use the clues to unscramble each word below. Write it in the spaces provided.

1. the sudden movement of rock and soil down a slope: DELINSDAL

___ ___ ___ ___ ___ ___ ___ ___ ___
(1)

2. the rapid mass movement of rock down a cliff: KOCR LAFL

___ ___ ___ ___   ___ ___ ___ ___
(6)

3. The traveling of a mass of material downslope is called mass _____.: NOTEMEMV

___ ___ ___ ___ ___ ___ ___ ___
(3)

4. the rapid movement of a large mass of mud: WOMFULD

___ ___ ___ ___ ___ ___ ___
(7)

5. steepest angle at which loose material no longer moves downslope: LEGNA FO EESPOR

___ ___ ___ ___ ___   ___ ___
(2) (4)

___ ___ ___ ___ ___
(9)

6. the most common type of landslide: PUMSL

___ ___ ___ ___ ___
(8)

7. the slow downhill movement of weathered rock material: PCEER

___ ___ ___ ___ ___
(5)

Write the numbered letters in the blanks below to find a word that describes mass movements.

8. ___ ___ ___ ___ ___ ___ ___ ___ ___
1 2 3 4 5 6 7 8 9

SECTION SUMMARY

Read the following section summary.

- Gravity causes rocks and soil to move downslope.
- If the slope on which material rests is greater than the angle of repose, mass movement will occur.
- Four types of mass movement are rock falls, landslides, mudflows, and creep.
- Landslides may destroy buildings and change wildlife habitats.

Name ______________________ Class ______________ Date ______________

Skills Worksheet

Directed Reading B

Section: The Active River (pp. 366–373)

RIVERS: AGENTS OF EROSION

1. How long ago was the area now known as the Grand Canyon nearly flat?

____

2. Wind, water, ice, or gravity moves soil and sediment from one location to another through a process called ______________________.

3. What has caused the Grand Canyon to be 1.6 km deep and 446 km long?

____

THE WATER CYCLE

______ **4.** What is the continuous movement of water between the atmosphere, the land, and the oceans?
a. erosion
b. evaporation
c. the water cycle
d. a river system

Match the correct description with the correct term. Write the letter in the space provided.

______ **5.** Water from the oceans and Earth's surface changes into water vapor; water gains energy.

______ **6.** Water flows over land into streams, rivers, and later enters oceans.

______ **7.** Rain, snow, sleet, or hail falls from clouds.

______ **8.** Water vapor cools and changes into water droplets that form clouds; water loses energy.

______ **9.** Water moves downward through spaces in soil due to gravity.

a. evaporation
b. percolation
c. condensation
d. precipitation
e. runoff

RIVER SYSTEMS

______ **10.** A network of streams and rivers that drains an area of its runoff is called
- **a.** a divide.
- **b.** a tributary.
- **c.** a water cycle.
- **d.** a river system.

11. A stream that flows into a lake or larger stream is a(n)

______________________.

12. The area of land that is drained by a water system is called a(n)

______________________.

13. What is the largest watershed in the United States?

____

14. The boundary between drainage areas that have streams that flow in opposite directions is a(n) ______________________.

STREAM EROSION

15. The path that a stream follows is a(n) ______________________.

Match the correct definition with the correct term. Write the letter in the space provided.

______ **16.** the measure of the change in elevation over a certain distance

______ **17.** the amount of water a stream or river carries in a given amount of time

______ **18.** the materials carried by a stream

- **a.** discharge
- **b.** load
- **c.** gradient

19. What effect does the gradient of a stream have on the amount of energy it has for eroding soil and rock?

____

____

____

____

20. When a stream's discharge increases, what happens to its erosive energy?

____

____

21. What effect does the speed of a flowing stream have on the size of the particles it is able to carry?

____

____

22. What effect does the size of the particles that make up a stream's load have on its erosive energy?

____

____

23. In a stream, what is a bed load?

____

____

____

24. In a stream, what is a dissolved load?

____

____

25. In a stream, what is a suspended load?

____

____

____

DESCRIBING RIVERS

______ **26.** The channel of a youthful river is
- **a.** deeper rather than wider.
- **b.** shallower rather than deeper.
- **c.** wider rather than deeper.
- **d.** deeper rather than shallower.

______ **27.** A youthful river has few tributaries and a gradient that is
- **a.** low.
- **b.** flat.
- **c.** curved.
- **d.** steep.

______ **28.** The channel of a mature river, compared to that of a youthful river, is
a. the same.
b. shallower rather than wider.
c. deeper rather than wider.
d. wider rather than deeper.

______ **29.** The gradient of a mature river, compared to that of a youthful river, is
a. the same.
b. steeper.
c. lower.
d. hard to measure.

______ **30.** A mature river has more discharge than a youthful river because
a. a mature river is longer.
b. a mature river is shorter.
c. it has a large watershed.
d. it has a small watershed.

______ **31.** A river with a low gradient, a wide, flat floodplain, and many bends is called
a. a youthful river.
b. a mature river.
c. an old river.
d. a rejuvenated river.

______ **32.** A river that forms where the land is raised by tectonic activity is
a. youthful.
b. mature.
c. old.
d. rejuvenated.

______ **33.** The steplike formations that often form on both sides of a rejuvenated river are called
a. meanders.
b. terraces.
c. stages.
d. gradients.

______ **34.** An old river deposits rock and soil
a. on its terraces.
b. in its tributaries.
c. along its channel.
d. across its watershed.

______ **35.** Fewer tributaries and an oxbow lake are common features of
a. a youthful river.
b. a mature river.
c. an old river.
d. a rejuvenated river.

Name ______________________ Class ______________ Date ________

Skills Worksheet

Directed Reading B

Section: Stream and River Deposits (pp. 374–377)

1. What bodies of water erode and move enormous amounts of soil and rock?

____

STREAM DEPOSITS

______ **2.** What happens to the load of rock and soil that rivers erode?

a. The load is picked up downstream.
b. The load is deposited downstream.
c. The load is dropped upstream.
d. The load is deposited upstream.

3. What is *deposition?*

____

____

4. Rocks and soil that streams deposit are called ______________.

5. Where do rivers and streams deposit sediment?

____

6. At what point does a river's current slow down?

____

____

7. What do the mud deposits of the deltas form?

____

____

8. In addition to causing the coastline to grow, what can deltas provide?

____

9. The Mississippi Delta formed from mud particles that came from places far

______________________.

10. What happens to the speed of a fast-moving mountain stream when it flows onto a flat plain?

____

11. In California's Death Valley, a fan-shaped mass of material deposited by a stream when the slope of the land decreases sharply is a(n) ______________________.

12. What is one difference between an alluvial fan and a delta?

____

____

FLOODS

13. What effect can flooding have on a stream?

____

14. An area along a river that forms from sediments deposited when the river overflows its banks is a(n) ______________________.

15. Why are floodplains rich farming areas?

____

____

16. What are three kinds of damage that floods can cause?

____

____

____

17. What is a barrier that can redirect and hold some of the floodwater in a reservoir called?

____

18. What is a buildup of sediment deposited along the channel of a river that keeps a river inside its banks?

____

19. What do people sometimes use to build artificial levees to control water during flooding?

____

Name ______________________ Class ______________ Date ____________

Skills Worksheet

Directed Reading B

Section: Using Water Wisely (pp. 378–383)

1. What percentage of Earth's water is drinkable?

____

2. How much of Earth's drinkable water is frozen in polar ice caps?

____

GROUNDWATER

_____ **3.** The water found inside rocks below Earth's surface is called
- **a.** the water table.
- **b.** an aquifer.
- **c.** an overdraft.
- **d.** groundwater.

_____ **4.** What is the rock layer that stores groundwater and allows the flow of groundwater called?
- **a.** the water table
- **b.** an aquifer
- **c.** an overdraft
- **d.** groundwater

_____ **5.** What is the upper surface of underground water called?
- **a.** the water table
- **b.** an aquifer
- **c.** an overdraft
- **d.** groundwater

6. What determines if the water table rises or falls?

____

WATER IN CALIFORNIA

7. What does it mean if an aquifer has an overdraft?

____

8. In California, surface water, such as rain, is captured and stored

in ______________________.

9. What part of the state receives about 75% of California's annual precipitation?

____

10. What part of California has about 75% of California's demand for water?

____

11. Why does California depend on a system that moves water from one part of the state to another?

____

12. Why does California receive water from nearby states, such as Oregon and Colorado?

____

13. List three major uses of California's water supply.

____

____

____

14. What kind of water use is urban use?

____

WATER POLLUTION

______ **15.** Waste matter or other material that is introduced into water and that is harmful to organisms is called

a. water deposition.
b. water pollution.
c. water porosity.
d. water permeability.

16. What are three sources of surface water pollution?

____

____

____

17. What is one thing people need to understand about pollutants in order to prevent pollution?

____

18. What law did the burning of the Cuyahoga River in 1969 help to pass?

____

19. Since the Clean Water Act of 1972 was passed, what has happened to the number of lakes and rivers that are fit for swimming and fishing?

____

20. List two other laws that protect the water supply.

____

____

WATER CONSERVATION

21. What is *conservation?*

____

____

______ **22.** One way that farmers can prevent water loss from evaporation and runoff is by
- **a.** recycling cooling water.
- **b.** installing low-flow toilets.
- **c.** using drip irrigation.
- **d.** using a hose with a shutoff nozzle.

______ **23.** What is one way industries can conserve used water, instead of discharging it into a nearby river?
- **a.** by recycling cooling water
- **b.** by installing low-flow toilets
- **c.** by using drip irrigation
- **d.** by using a hose with a shutoff nozzle

______ **24.** Households can conserve water in the bathroom by
- **a.** recycling cooling water.
- **b.** installing low-flow toilets.
- **c.** using drip irrigation.
- **d.** using a hose with a shutoff nozzle.

______ **25.** What is xeriscaping?
- **a.** landscaping with exotic plants
- **b.** landscaping with houseplants
- **c.** landscaping with edible plants
- **d.** landscaping with native plants

______ **26.** You can conserve water when washing a car by using a bucket and
- **a.** recycling cooling water.
- **b.** installing low-flow toilets.
- **c.** using drip irrigation.
- **d.** using a hose with a shutoff nozzle.

Name ______________________ Class ______________ Date ____________

Skills Worksheet

Vocabulary and Section Summary B

The Active River

VOCABULARY

After you finish reading the section, try this puzzle! Use the clues below to solve the following crossword puzzle.

ACROSS

5. A river causes _____ by removing rocks and soil from its riverbed.

6. An area of land drained by a water system is called a(n) _____.

7. The continuous movement of water from lakes and oceans into the air, onto land, into the ground, and back to lakes is called the _____.

DOWN

1. Drainage basins are separated from each other by an area of higher ground called a(n) _____.

2. The path that a stream follows is called a(n) _____.

3. A small river that flows into a larger river is called a(n) _____.

4. The term for the materials carried by a stream is _____.

SECTION SUMMARY

Read the following section summary.

- Rivers shape Earth's landscape through the process of erosion.
- The sun is the major source of energy that drives the water cycle.
- A river system is made up of a network of streams and rivers.
- A watershed is a region that collects runoff water that then becomes part of a river system that drains into a lake or the ocean.
- Gradient affects the speed at which water flows in a stream. The higher the gradient, the faster the water flows. Water that flows rapidly has a lot of energy for eroding soil and rock.
- When a stream's discharge increases, the stream's erosive energy also increases.
- A stream can carry eroded particles as bed load, suspended load, or dissolved load. A stream that has a load of large particles has a high rate of erosion.
- A river can be described as youthful, mature, old, or rejuvenated based on its characteristics.

Name ______________________ Class ______________ Date ____________

Skills Worksheet

Vocabulary and Section Summary B

Stream and River Deposits

VOCABULARY

After you finish reading the section, try this puzzle! In the space provided, write the term described. Then, find the words in the word search puzzle on the next page. Terms can be hidden in the puzzle vertically, horizontally, diagonally, or backward.

______________ **1.** the process in which material is laid down

______________ **2.** a fan-shaped mass of material deposited by a stream when the slope of the land decreases sharply

______________ **3.** a buildup of sediment deposited along the channel of a river

______________ **4.** an area along a river that forms from sediments deposited when the river overflows its banks

______________ **5.** rock and soil deposited by streams

______________ **6.** a barrier that can redirect and hold a portion of the floodwater in a reservoir

______________ **7.** a fan-shaped mass of material deposited at the mouth of a stream

Vocabulary and Section Summary B *continued*

O	G	G	D	Z	B	X	E	D	N	E	N	X	I	Q
M	A	T	E	T	R	A	S	O	O	P	A	P	C	D
Z	B	A	L	E	M	A	I	D	B	H	F	Q	O	I
Y	H	E	T	F	R	T	U	X	C	W	L	P	U	S
D	N	U	A	N	I	R	I	P	U	Z	A	K	S	R
X	H	I	E	S	S	L	F	P	I	N	I	Y	E	T
Y	A	S	O	G	E	Y	S	G	S	Y	V	R	S	N
I	P	P	E	V	E	U	Y	W	H	D	U	D	O	M
U	E	U	E	D	F	L	O	O	D	P	L	A	I	N
D	J	E	N	T	I	D	D	H	L	H	L	J	X	U
V	B	Y	M	A	D	M	J	E	X	Z	A	U	Q	T
B	E	O	I	I	N	R	E	M	B	D	M	D	W	R
L	B	B	J	R	N	L	E	N	O	F	F	T	D	X
Z	L	U	C	L	A	Z	B	B	T	Z	I	W	O	W

SECTION SUMMARY

Read the following section summary.

- Sediment forms several types of deposits, such as deltas, alluvial fans, and floodplains.
- A delta is a fan-shaped deposit of sediment that forms where a river meets a large body of water.
- Alluvial fans can form when a river deposits sediment on land.
- Flooding brings rich soil to farmland and may cause a stream to change course.
- Flooding can also lead to property damage and death.

Name ______________________ Class ______________ Date ____________

Skills Worksheet

Vocabulary and Section Summary B

Using Water Wisely

VOCABULARY

After you finish reading the section, try this puzzle! Use the clues below to unscramble the letters, and write the word in the space provided.

1. type of landscaping that uses native plants: GEISRCXAPIN

____

2. the upper surface of underground water: TEWAR LEATB

____

3. the preservation and wise use of natural resources: NOITSEAVRONC

____

4. a body of rock or sediment that stores groundwater and allows it to flow: QUAFREI

____

5. process by which more water is being pumped out of an aquifer than is being replaced by rain: FREVDORAT

____

6. the water found inside the rocks below Earth's surface: RRTGONAUDWE

____

7. waste matter or other material that is introduced into water and is harmful to organisms that live in, drink, or are exposed to the water: RETWA LUNIOTLPO

____

SECTION SUMMARY

Read the following section summary.

- An aquifer is a rock and soil layer that stores groundwater and allows the flow of groundwater.
- California receives its water from surface water, from aquifers, and from other areas.
- Water sources can be polluted by cities, factories, and farms.
- Water can be conserved by using only the water that is needed, by recycling water, and by using drip irrigation systems.

Name ______________________ Class ______________ Date ______________

Skills Worksheet

Directed Reading B

Section: Earth's Oceans (pp. 400–406)

1. What percentage of Earth's surface is covered with water?

____

2. The global ocean is divided by the continents into ______________ main oceans.

DIVISIONS OF THE GLOBAL OCEAN

Match the correct definition with the correct term. Write the letter in the space provided.

______ **3.** the largest ocean

______ **4.** the ocean that is about half the volume of the Pacific Ocean

______ **5.** the third-largest ocean

______ **6.** the ocean whose surface is partly covered by ice

______ **7.** the ocean that extends from the coast of Antarctica to 60° south latitude

a. Atlantic Ocean
b. Arctic Ocean
c. Indian Ocean
d. Southern Ocean
e. Pacific Ocean

CHARACTERISTICS OF OCEAN WATER

______ **8.** Which of the following statements about salt in the ocean is true?
a. It is sodium chloride.
b. It is poisonous.
c. It is saltier than the salt we eat.
d. It is less salty than the salt we eat.

______ **9.** The measure of the amount of dissolved salts in a given amount of liquid is called
a. circulation.
b. sodium chloride.
c. salinity.
d. evaporation.

______ **10.** Water salinity is usually higher in
a. coastal waters in cool, humid environments.
b. river waters.
c. coastal waters in hot, dry climates.
d. coastal waters near river outlets.

______ **11.** When water evaporates,
a. dissolved salts in the water also evaporate.
b. dissolved salt breaks down into its base elements.
c. dissolved salts rise to the surface of the water.
d. dissolved salts in the water are left behind.

______ **12.** Which of the following does NOT affect ocean salinity?
a. climate
b. sea animals
c. inflow of fresh water
d. water movement

13. How do minerals on land make oceans salty?

____

____

____

TEMPERATURE OF OCEAN WATER

Match the correct definition with the correct term. Write the letter in the space provided.

______ **14.** second, cooler layer of ocean water

______ **15.** warm, top layer of ocean water

______ **16.** bottom, coolest layer of ocean water

a. surface zone
b. thermocline
c. deep zone

17. In the ______________________, temperature drops with greater depth faster than it does in the other two zones.

18. What two things affect the surface temperatures of oceans in most regions?

____

____

19. What two things affect the density of ocean water?

____

____

Name ____________________ Class ____________ Date ____________

Skills Worksheet

Directed Reading B

Section: The Ocean Floor (pp. 406–411)

STUDYING THE OCEAN FLOOR

______ **1.** *Sonar* stands for
- **a.** sound and radar.
- **b.** sound navigation and radio.
- **c.** sub-ocean navigation and ranging.
- **d.** sound navigation and ranging.

______ **2.** *Geosat* measures
- **a.** changes in the height of ocean surface.
- **b.** changes in the ocean's landforms.
- **c.** the direction and speed of ocean currents.
- **d.** the velocity of waves hitting the shoreline.

______ **3.** Using satellites to make maps of the ocean floor is better than using ship-based sonar because
- **a.** satellites are farther away from the ocean.
- **b.** satellites cover more territory.
- **c.** ships are always getting lost.
- **d.** satellites are less expensive.

4. How deep will future models of *Deep Flight* take scientists?

____

____

____

5. What is an advantage of using *Jason II* to study the ocean?

____

____

____

6. How fast does sound travel in water?

____

____

7. How do scientists use sound to figure out the depth of the ocean?

____

____

OCEAN FLOOR BASICS

8. The two major regions of land under the water are the continental margin and the ______________________ basin.

9. If the ocean were a giant swimming pool, the ______________________ would be the shallow end.

10. What are the three subdivisions of the continental margin?

____

____

____

11. What does the deep-ocean basin consist of?

____

____

____

____

TECTONIC PLATES AND OCEAN-FLOOR FEATURES

12. What forms features on the ocean floor?

____

13. The flattest regions on Earth are called ______________________.

14. Mountains on the ocean floor that can turn into volcanic islands are called ______________________.

15. Mountain chains formed by magma coming through rift zones are called ______________________.

16. When plates move apart from each other, the motion causes a crack in the ocean floor called a(n) ______________________.

17. When one oceanic plate is forced underneath another plate, a(n) ______________________ forms.

18. Ocean trenches are formed by the process called ______________________.

Name ______________________ Class ______________ Date ______________

Skills Worksheet

Directed Reading B

Section: Resources from the Ocean (pp. 412–415)

1. List three resources available from the ocean.

____

____

____

LIVING RESOURCES

2. Fish are considered a(n) ______________________ resource.

3. Fish populations are reduced by ______________________.

4. What kinds of seafood are raised and harvested in farms?

____

____

5. What is kelp?

____

6. What are some ways we use seaweed?

____

____

NONLIVING RESOURCES

7. What is desalination?

____

____

____

8. Name three advantages of tidal energy.

____

____

____

____

9. How can water during a high tide be used to generate electricity?

____

____

____

10. Oil and natural gas are ______________ resources.

11. How do petroleum workers use readings about the ocean floor from seismic devices?

____

____

____

12. Mineral nodules in the ocean are made mostly of ______________.

13. What are two minerals that make up nodules?

____

____

Name ______________________ Class ______________ Date ____________

Skills Worksheet

Directed Reading B

Section: Ocean Pollution (pp. 416–421)

______ **1.** Which of the following statements is NOT true?
a. Humans throw trash in the oceans.
b. Humans have thrown trash into the oceans for hundreds, if not thousands, of years.
c. Humans are becoming more aware of ocean pollution.
d. Trash in the oceans is not harmful to plants, animals, or people.

NONPOINT-SOURCE POLLUTION

______ **2.** Nonpoint-source pollution
a. is easily identified.
b. comes from one specific source.
c. is less harmful than other kinds of pollution.
d. comes from many sources.

______ **3.** Human activities on land can pollute streams and rivers, which then flow into the ocean and
a. clean it up.
b. pollute it.
c. produce kelp.
d. feed marine animals.

4. Briefly explain two ways people contribute to nonpoint-source pollution.

____

____

POINT-SOURCE POLLUTION

______ **5.** Point-source pollution
a. comes from unidentified sources.
b. comes from a specific source.
c. is harmful only to plants.
d. is not a problem today.

______ **6.** What is NOT an example of point-source pollution?
a. trash washed up on beaches
b. treated sludge
c. pesticides from people's lawns
d. oil spills

_____ **7.** What is sludge?

a. the solid part of waste matter
b. oil spilled from tankers
c. plastic trash
d. waste gasoline

8. How is plastic trash harmful to marine animals?

____

____

____

9. What are two effects of oil spills?

____

____

10. How are oil tankers being built to prevent oil spills?

____

____

SAVING OUR OCEAN RESOURCES

11. In 1989, sixty-four countries ratified a treaty that ______________ dumping of certain metals, plastics, oils, and radioactive wastes into the ocean.

12. One of the largest beach clean-up programs in the United States is the semiannual ______________ program.

13. Volunteer organizations have cleaned up millions of tons of trash, and they've also helped educate people about the ______________ of ocean dumping.

14. In 1972, Congress passed the ______________, which put the Environmental Protection Agency in charge of issuing permits for dumping trash into the ocean.

15. Briefly describe what the U.S. Marine Protection, Research, and Sanctuaries Act does.

____

____

Name ______________________ Class ______________ Date ______________

Skills Worksheet

Vocabulary and Section Summary B

Earth's Oceans

VOCABULARY

After you finish reading the section, try this puzzle! Use the clues below to solve the crossword puzzle on the following page.

ACROSS

4. This is the zone directly below the thermocline. (two words)

6. The third largest ocean is the _____ Ocean.

8. The ocean that is covered with ice is the _____ Ocean.

9. The ocean that extends from Antarctica to 60° south latitude is the _____ Ocean.

10. This is the top layer of ocean water. (two words)

DOWN

1. The second largest body of water on Earth is the _____ Ocean.

2. This is the zone in which water temperature drops more rapidly with depth.

3. This is the chemical name for sea salt. (two words)

5. The largest body of water on Earth is the _____ Ocean.

7. This is the amount of dissolved salts in a liquid.

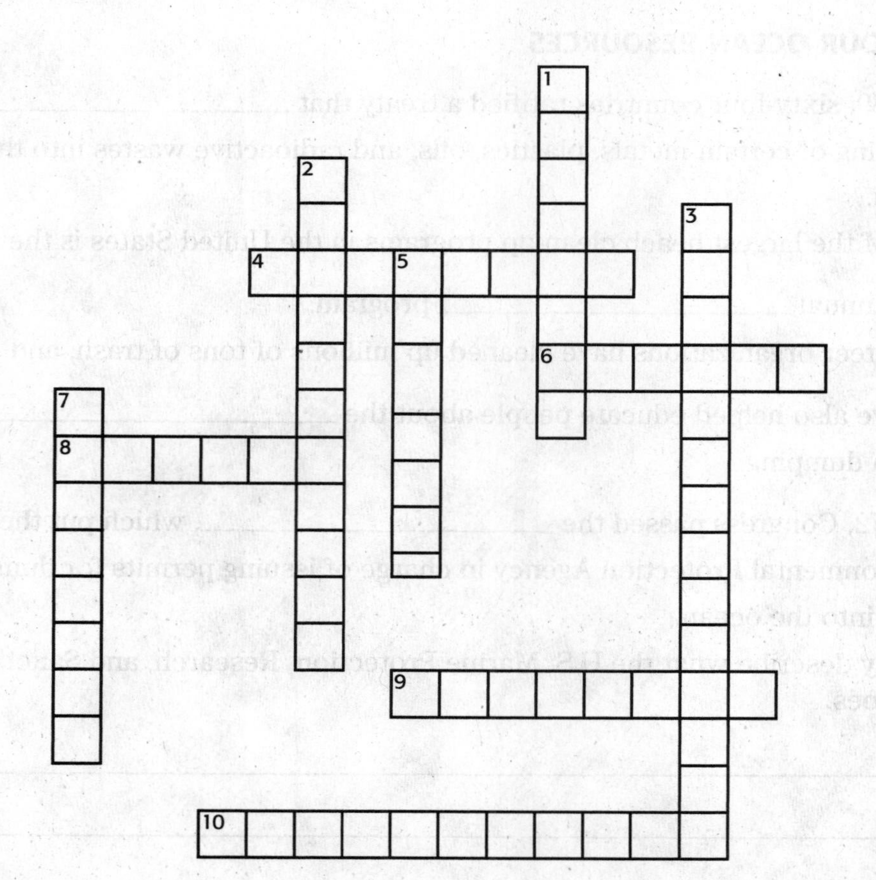

SECTION SUMMARY

Read the following section summary.

- The global ocean is divided by the continents into five main oceans: Pacific Ocean, Atlantic Ocean, Indian Ocean, Southern Ocean, and Arctic Ocean.
- Salts have collected in the ocean for billions of years. Salinity is a measure of the amount of dissolved salts in a given mass of liquid.
- The three temperature zones of ocean water are the surface zone, the thermocline, and the deep zone.
- Temperature and salinity determine the density of ocean water. The density of ocean water drives convection currents.

Name ______________________ Class ______________ Date ______________

Skills Worksheet

Vocabulary and Section Summary B

The Ocean Floor

VOCABULARY

After you finish reading the section, try this puzzle! The underlined words below are scrambled. Write the completed words in the spaces provided.

1. Scientists use ARNOS to measure the ocean's depth by sending sound pulses down into the ocean.

   ____

2. When one tectonic plate moves under another it is called DUCSBUTONI.

   ____

3. A(n) YUTGO is a flat-topped, submerged seamount.

   ____

4. The flattest regions on Earth are called BALSYSA SLAPIN.

   ____

5. A(n) DIM-EANCO DIRGE is a long, undersea mountain chain along the ocean's floor.

   ____

6. A(n) SONUTAME is a submerged volcanic mountain on the ocean floor.

   ____

7. A(n) EANCO RENTCH is formed by subduction.

   ____

SECTION SUMMARY

Read the following section summary.

- Scientists study the ocean floor by using sonar, underwater vessels, drilling, and satellites.
- The ocean floor is divided into two regions—the continental margin and the deep-ocean basin.
- The continental margin consists of the continental shelf, the continental slope, and the continental rise.
- The deep-ocean basin consists of the abyssal plain, mid-ocean ridges, trenches, and seamounts.
- Mid-ocean ridges and trenches mark the boundaries of tectonic plates.

Name ______________________ Class ______________ Date ____________

Skills Worksheet

Vocabulary and Section Summary B

Resources from the Ocean

VOCABULARY

After you finish reading the section, try this puzzle! Use the clues below to find the correct term. Then, find those words in the puzzle on the following page. Terms can be hidden in the puzzle vertically, horizontally, or backward.

______________ **1.** This is a form of seaweed that is often used as a thickener in jellies and ice cream.

______________ **2.** This is generated by the movement of the tides at La Rance off the coast of France. (two words)

______________ **3.** This is the process of removing salt from ocean water.

______________ **4.** These form when dissolved minerals, such as manganese, form around solid objects.

______________ **5.** This is one of the two major regions of ocean along which offshore oil and natural gas deposits are found. (two words)

Vocabulary and Section Summary B *continued*

S	Y	G	Q	E	V	Y	T	Z	A	H	Z	Y	H	Z	V	S
F	D	P	L	D	A	D	K	S	T	Q	B	E	L	V	B	E
A	T	H	E	E	J	K	J	E	Y	Z	E	H	C	P	H	A
J	H	A	T	S	P	L	A	G	E	H	A	D	A	D	C	H
Z	Q	H	Y	A	C	Z	Y	D	J	E	U	E	O	K	L	Y
Q	N	E	E	L	H	S	Q	X	Q	F	Y	S	B	Y	D	T
C	O	N	T	I	N	E	N	T	A	L	M	A	R	G	I	N
Y	D	A	C	N	H	T	H	D	J	Y	T	L	K	B	A	D
S	U	E	Q	A	D	Z	T	E	H	E	D	I	P	L	E	K
G	L	X	S	T	A	V	S	V	Z	B	T	N	L	Z	K	S
E	E	T	T	I	D	A	L	E	N	E	R	G	Y	A	P	H
A	S	H	E	O	D	T	H	G	V	E	V	A	K	C	H	T
Y	K	N	A	N	V	E	R	P	T	Y	J	S	B	V	L	Y
Q	D	C	S	F	Z	A	X	E	D	Z	Q	T	H	C	Z	A

SECTION SUMMARY

Read the following section summary.

- Humans depend on the ocean for living and nonliving resources.
- Fish and other marine life are caught in the ocean and are being raised in fish farms to help feed growing human populations.
- Nonliving ocean resources include oil and natural gas, fresh water, minerals, and tidal energy.

Name ______________________ Class ______________ Date ____________

Skills Worksheet

Vocabulary and Section Summary B

Ocean Pollution

VOCABULARY

After you finish reading the section, try this puzzle! Use the clues given to fill in the blanks below. Then, copy the numbered letters into the corresponding boxes on the following page to reveal the answer to the bonus question.

1. the second largest ocean in which there was a major oil spill in 1988

___ ___ ___ ___ ___ ___ ___ ___
4 7 12

2. the solid waste of raw sewage

___ ___ ___ ___ ___ ___
2

3. the type of pollution that comes from a specific site

___ ___ ___ ___ ___ - ___ ___ ___ ___ ___ ___
8 9

4. the type of pollution that comes from many sources

___ ___ ___ ___ ___ ___ ___ ___ - ___ ___ ___ ___ ___ ___
5 13 1

5. all the liquid and solid wastes that are flushed down toilets and poured down drains

___ ___ ___ ___ ___ ___ ___ ___ ___
10 11 3 6

BONUS QUESTION

6. What law put the Environmental Protection Agency in charge of regulating the dumping of trash in the ocean?

___ ___ ___ ___ ___ ___ ___ ___ ___ ___ ___ ___ ___
1 2 3 4 5 6 7 8 9 10 11 12 13

SECTION SUMMARY

Read the following section summary.

- The two main types of water pollution are nonpoint-source pollution and pointsource pollution.
- Types of nonpoint-source pollution include oil and gasoline from cars, trucks, and watercraft, as well as of pesticides, herbicides, and fertilizers.
- Oil spills harm wildlife and local fishing economies and cost billions of dollars to clean up.
- Efforts to save ocean resources include laws, international treaties, and volunteer cleanups.

Name ______________________ Class ______________ Date ______________

Skills Worksheet

Directed Reading B

Section: Currents (pp. 436–443)

______ **1.** Streamlike movements of water are called
- **a.** ocean currents.
- **b.** the Coriolis effect.
- **c.** global circulation.
- **d.** continental deflections.

2. Name three factors that influence currents.

____

____

____

SURFACE CURRENTS

3. Horizontal movements of water that are caused by wind and occur at or near the ocean's surface are called ______________.

4. Name three factors that control surface currents.

____

____

____

5. Which way do the winds blow near the equator?

____

6. Explain how the sun is the major source of energy that powers surface currents.

____

____

____

7. What causes some wind and ocean currents to be deflected from the paths they would take?

____

8. What is the deflection of moving objects from a straight path due to Earth's rotation called?

____

9. What happens to surface currents when the currents meet a continent?

____

____

10. What is the transfer of energy as a result of the movement of matter called?

____

DEEP CURRENTS

Match the correct description with the correct term. Write the letter in the space provided.

______	**11.** movement of ocean water located far below the surface	**a.** density
______	**12.** the amount of matter in a given space or volume	**b.** deep current
______	**13.** a measure of the amount of dissolved salts or solids in a liquid	**c.** salinity

14. What happens to ocean water as it becomes denser?

____

____

15. What is the deepest and densest layer of water in the ocean?

____

____

CONVECTION CURRENTS

16. Why are surface currents and deep currents also called convection currents?

____

____

GLOBAL CIRCULATION

17. Name three materials transported by global ocean circulation.

____

____

Name ______________________ Class ______________ Date ____________

Skills Worksheet

Directed Reading B

Section: Currents and Climate (pp. 444–447)

SURFACE CURRENTS AND CLIMATE

______ **1.** Coastal areas that have climates that are warmer than would be expected at their latitude are affected by
- **a.** cold-water currents.
- **b.** warm-water currents.
- **c.** deep currents.
- **d.** breaking currents.

______ **2.** Where does the Gulf Stream get its warmth?
- **a.** from the Tropics
- **b.** from the North Atlantic
- **c.** from the British Isles
- **d.** from Newfoundland

______ **3.** The California Current carries cold water south from
- **a.** the South Atlantic Ocean.
- **b.** the South Pacific Ocean.
- **c.** the North Atlantic Ocean.
- **d.** the North Pacific Ocean.

______ **4.** The climate along the West Coast of the United States is cooler than the climate of inland areas because of the effects of
- **a.** cold-water currents.
- **b.** warm-water currents.
- **c.** deep currents.
- **d.** breaking currents.

EL NIÑO AND LA NIÑA

5. In what ocean do El Niño and La Niña form?

____

6. What is *El Niño*?

____

____

7. What is *La Niña*?

____

____

____

8. What do scientists need to study in order to predict the weather changes on land that might be caused by El Niño?

____

____

____

9. Why is it important for scientists to learn as much as possible about El Niño?

____

____

____

10. Explain one way that scientists collect data to predict El Niño.

____

____

11. What data do buoys record?

____

____

12. Describe two weather patterns that El Niño can alter that can cause disasters.

____

____

____

____

UPWELLING

______ **13.** Cold, nutrient-rich water from the deep ocean rises to the surface and replaces warm surface water in a process called

a. the Coriolis effect.
b. upwelling.
c. El Niño.
d. La Niña.

14. Give one reason why upwelling is important to ocean life.

____

____

____

Name ______________________ Class ______________ Date ______________

Skills Worksheet

Directed Reading B

Section: Waves and Tides (pp. 448–455)

______ **1.** What is a disturbance that transmits energy through matter or empty space?
- **a.** evaporation
- **b.** trough
- **c.** deflection
- **d.** wave

WAVES

2. The highest part of a wave is the ______________________.

3. The lowest part of a wave is the ______________________.

4. The distance between two adjacent wave crests or troughs is called the ______________________.

5. The vertical distance between the crest and the trough of a wave is the ______________________.

6. How do most ocean waves form?

____

____

____

7. What happens to water when a wave of energy moves through it?

____

____

8. Why are waves different sizes?

____

____

9. What is the time between the passage of two wave crests (or troughs) at a fixed point?

____

____

10. What does dividing wavelength by wave period give you?

____

11. When waves reach the shore, what is transferred to the beach environment?

____

12. Describe how a breaker forms.

____

____

____

TIDES

13. What are tides?

____

____

14. What two forces influence tides?

____

____

____

15. Why is the gravitational pull on liquids more noticeable than the pull on solids?

____

____

16. What does the pull of the moon on the part of the ocean that directly faces the moon cause?

____

17. What are high tides?

____

____

18. Where do low tides form?

____

____

19. How long does it take for a spot on Earth that is facing the moon to rotate so that it is facing the moon again?

____

TIDAL VARIATIONS

20. The difference between levels of ocean water at high tide and low tide is called a(n) ______________.

21. Tides that occur when the sun, Earth, and the moon are aligned are called ______________.

22. Tides that occur when the sun, Earth, and the moon form a 90° angle are called ______________.

Name ______________________ Class ______________ Date ____________

Skills Worksheet

Vocabulary and Section Summary B

Currents

VOCABULARY

After you finish reading the section, try this puzzle! Fill in each blank with the correct term. Two blanks indicate a two-word term. Then, circle the words in the puzzle on the next page. Words may appear horizontally, vertically, or diagonally.

1. The curving of the path of a moving object from an otherwise straight path due to Earth's rotation is called the ______________________ ______________________.

2. A streamlike movement of ocean water located far below the surface is a(n) ______________________ ______________________.

3. Any movement of matter that results from differences in density is called a(n) ______________________ current.

4. The water that flows deep in the ocean has more ______________________ than water at the ocean surface.

5. Global winds, the Coriolis effect, and continental deflections keep ______________________ ______________________ flowing in distinct patterns around Earth.

6. The measure of the amount of dissolved salts or solids in a liquid is its ______________________.

C	R	S	U	E	A	O	C	T	D	F	E	E	S	S
O	T	C	L	D	M	N	P	C	U	R	I	L	H	U
R	W	G	O	E	T	C	V	E	C	T	F	C	G	R
I	S	U	R	N	E	I	L	A	N	B	R	E	S	F
O	L	O	E	S	V	S	I	O	L	I	C	U	A	A
L	F	G	L	I	A	E	S	T	R	R	F	R	L	C
I	H	B	E	T	Y	I	C	A	R	N	B	E	I	E
S	I	A	O	Y	S	U	R	T	C	E	O	P	N	C
E	O	J	K	O	M	N	S	A	I	U	R	R	I	U
F	E	N	T	I	E	E	P	A	T	O	E	D	T	R
F	S	L	I	N	Y	L	L	S	D	R	N	R	Y	R
E	D	E	E	P	C	U	R	R	E	N	T	G	J	E
C	R	I	Y	T	P	I	M	N	O	L	V	E	C	N
T	E	W	F	S	V	R	F	P	T	V	U	V	L	T
D	W	M	N	E	I	O	I	R	S	P	E	M	N	S

SECTION SUMMARY

Read the following section summary.

- Surface currents form as wind transfers energy to ocean water.
- Surface currents are controlled by three factors: global winds, the Coriolis effect, and continental deflections.
- Deep currents form where the density of ocean water increases. Water density depends on temperature and salinity.
- Surface currents and deep currents combine to form convection currents that transfer energy.
- Earth's global circulation moves water through all oceans and distributes materials and heat.

Name ______________________ Class ______________ Date __________

Skills Worksheet

Vocabulary and Section Summary B

Currents and Climate

VOCABULARY

After you finish reading the section, try this puzzle! Use the clues below to fill in the following crossword puzzle.

ACROSS

2. brings nutrients to the surface of the ocean

4. causes surface water temperatures to become unusually cool

5. long periods during which rainfall is below average

DOWN

1. currents that greatly affect the climate

3. can alter weather patterns enough to cause disaster

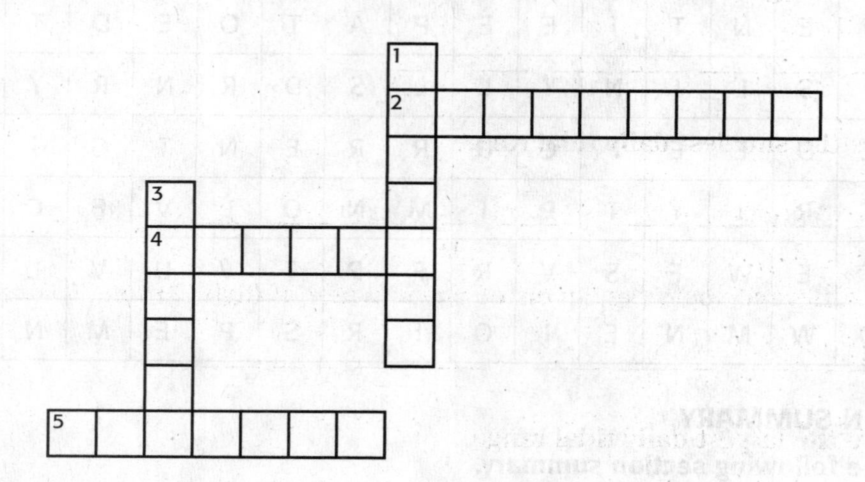

SECTION SUMMARY

Read the following section summary.

- Surface currents cause climates of coastal areas to be more moderate than inland climates at the same latitude and elevation.
- Upwelling is the flow of cold, nutrient-rich water from the deep ocean to the surface.
- During El Niño, warm and cool surface waters change locations. El Niño can cause floods, mudslides, drought, and changes in upwelling.

Name ______________________ Class ______________ Date ______________

Skills Worksheet

Vocabulary and Section Summary B

Waves and Tides

VOCABULARY

After you finish reading the section, try this puzzle! Use the clues given to fill in the blanks below. Then, copy the numbered letters above the corresponding numbers to label the wave drawing.

1. a daily change in the level of ocean water

___ ___ ___ ___
4

2. any disturbance that transmits energy through matter or empty space

___ ___ ___ ___
6 8

3. when a high wave crest crashes down onto the ocean floor

___ ___ ___ ___ ___ ___ ___
1

4. have the smallest daily tidal range

___ ___ ___ ___ ___ ___ ___ ___ ___
2 11

5. the difference between levels of ocean water at high tide and low tide

___ ___ ___ ___ ___ ___ ___ ___ ___ ___
9 7 5

6. have the largest daily tidal range

___ ___ ___ ___ ___ ___ ___ ___ ___ ___
3 10

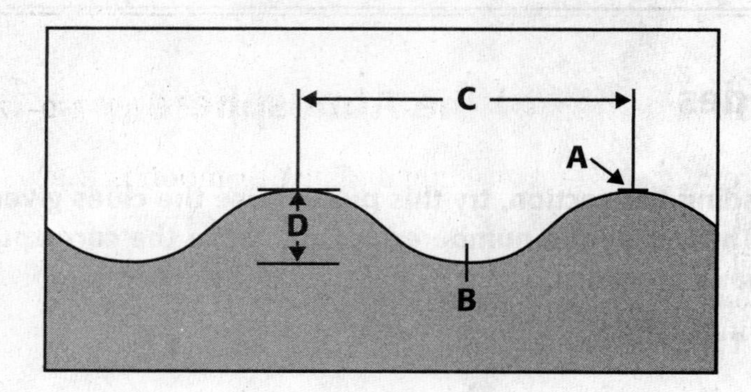

a. C ___ ___ ___ ___
1 2 3 4

b. ___ ___ O U ___ H
4 1 5

c. ___ ___ ___ ___ ___ ___ ___ ___ ___ H
6 7 8 2 9 2 10 5 4

d. ___ ___ ___ ___ H ___ ___ ___ H ___
6 7 8 2 2 11 5 4

SECTION SUMMARY

Read the following section summary.

- Waves form when the wind's energy is transferred to the surface of the ocean.
- Wave energy travels through water near the water's surface, while the water itself rises and falls in circular movements.
- Waves break when the water depth becomes so shallow that the bottom of the wave transfers energy to the ocean bottom and the shore.
- Tides are caused by the gravitational forces of the moon and the sun on Earth. The moon's gravity is the main force behind the tides.
- The positions of the sun and moon relative to the position of Earth cause tidal ranges.

Name ______________________ Class ______________ Date ____________

Skills Worksheet

Directed Reading B

Section: Characteristics of the Atmosphere (pp. 470–473)

______ **1.** A mixture of gases surrounding a planet or moon is the
- **a.** oxygen.
- **b.** atmosphere.
- **c.** breathable air.
- **d.** hemisphere.

THE COMPOSITION OF THE ATMOSPHERE

______ **2.** The most common atmospheric gas on Earth is
- **a.** oxygen.
- **b.** argon.
- **c.** nitrogen.
- **d.** carbon dioxide.

______ **3.** Living things, such as plants, produce the atmosphere's
- **a.** oxygen.
- **b.** argon.
- **c.** nitrogen.
- **d.** carbon dioxide.

______ **4.** Most water in the atmosphere is in
- **a.** rain.
- **b.** ice.
- **c.** water vapor.
- **d.** carbon dioxide.

AIR PRESSURE AND TEMPERATURE

______ **5.** At sea level, a square inch of surface area is under almost how many pounds of pressure?
- **a.** 150 lb
- **b.** 15 lb
- **c.** 30 lb
- **d.** 1500 lb

______ **6.** Gas molecules in the atmosphere are pulled toward Earth by
- **a.** air pressure.
- **b.** the moon.
- **c.** gravity.
- **d.** surface area.

7. The measure of the force with which air molecules push on a surface is called ______________________.

8. What happens to air pressure as you move away from Earth's surface?

____

____

9. Why are some parts of the atmosphere warmer than others?

____

____

LAYERS OF THE ATMOSPHERE

Match the correct definition with the correct term. Write the letter in the space provided.

_____ **10.** coldest layer of the atmosphere

_____ **11.** atmosphere layer including the ozone layer

_____ **12.** layer of atmosphere closest to Earth

_____ **13.** uppermost layer of the atmosphere

a. troposphere

b. mesosphere

c. stratosphere

d. thermosphere

14. How are the layers of the atmosphere defined?

____

____

15. In the troposphere and mesosphere, what happens to the temperature as altitude increases?

____

____

____

16. Why do the stratosphere and thermosphere have high temperatures?

____

____

____

Name ______________________ Class ______________ Date ____________

Skills Worksheet

Directed Reading B

Section: Atmospheric Heating (pp. 474–479)

______ **1.** What is the travel time for solar energy to reach Earth?
- **a.** about 8 hours
- **b.** about 80 hours
- **c.** about 8 minutes
- **d.** about 8 days

RADIATION: ENERGY TRANSFER BY WAVES

______ **2.** What percentage of the energy radiated by the sun reaches Earth's surface?
- **a.** two-fiftieths
- **b.** two-thousandths
- **c.** two-millionths
- **d.** two-billionths

______ **3.** What percentage of the sun's energy that reaches Earth is absorbed by Earth's surface?
- **a.** 25%
- **b.** 50%
- **c.** 20%
- **d.** 5%

______ **4.** What percentage of the sun's energy that reaches Earth is absorbed by ozone, clouds, and atmospheric gases?
- **a.** 25%
- **b.** 50%
- **c.** 20%
- **d.** 5%

5. The collection of all frequencies or wavelengths of electromagnetic waves is called ______________________.

6. Solar radiation is ______________________ by the different layers of Earth's atmosphere.

7. Most solar energy that reaches Earth's surface takes the form of ______________________.

CONDUCTION: ENERGY TRANSFER BY CONTACT

8. How is heat transferred in conduction?

____

____

____

____

9. Why is heat always transferred from warmer areas to colder areas?

____

____

____

____

CONVECTION: ENERGY TRANSFER BY MOTION

Match the correct description with the correct term. Write the letter in the space provided.

______ **10.** transfer of energy by circulation or movement of a liquid or gas

______ **11.** circular movement of warm fluid rising and cool fluid sinking

a. convection

b. convection current

THE GREENHOUSE EFFECT

12. Explain what process produces the greenhouse effect.

____

____

____

____

13. The balance between incoming solar energy and outgoing energy radiated into space is called ______________________ and makes Earth livable.

GLOBAL WARMING

14. A gradual increase in average global temperature is called

____________________.

15. What are greenhouse gases?

____

16. What human activities may increase the level of greenhouse gases in the atmosphere?

____

____

Name ______________________ Class ______________ Date ____________

Skills Worksheet

Directed Reading B

Section: Air Movement and Wind (pp. 480–483)

WHAT CAUSES WIND?

______ **1.** What causes differences in air pressure?
a. even heating of Earth
b. even cooling of Earth
c. uneven heating of Earth
d. increased heating of Earth

______ **2.** The movement of air caused by differences in air pressure is called
a. dense air.
b. wind.
c. polar air.
d. vents.

______ **3.** Air is warmer and less dense than surrounding air at the equator because the equator receives more
a. wind.
b. air pressure.
c. solar energy.
d. radiation.

______ **4.** Because air at the poles is colder and denser than surrounding air, it
a. rises.
b. sinks.
c. circulates.
d. stagnates.

______ **5.** High pressure areas are created around the poles as cold air
a. rises.
b. blows.
c. stagnates.
d. sinks.

6. When the paths of winds and ocean currents curve because of Earth's rotation, it's called the ______________________.

GLOBAL WINDS

Match the correct description with the correct term. Write the letter in the space provided.

______ **7.** winds that blow from 30° latitude in both hemispheres almost to the equator

______ **8.** winds formed when convection cells, pressure belts, and winds combine with the Coriolis effect

______ **9.** wind formed as cold, sinking air moves from the poles to 60° north and 60° south latitude

______ **10.** wind belts that extend between 30° and 60° latitude in both hemispheres

a. polar easterlies
b. westerlies
c. trade winds
d. global winds

LOCAL WINDS

11. How can the properties of matter making up Earth's surface cause local winds?

____

____

____

Match the correct description with the correct term. Write the letter in the space provided.

______ **12.** cool air that flows over the ocean toward the land during the day

______ **13.** cool air that flows over the land toward the ocean during the night

______ **14.** warm air that rises up mountain slopes during the day

______ **15.** cool air that moves down mountain slopes during the night

a. valley breeze
b. land breeze
c. sea breeze
d. mountain breeze

Name _______________ Class _______________ Date _______________

Skills Worksheet

Directed Reading B

Section: The Air We Breathe (pp. 484–491)

AIR POLLUTION

1. The contamination of the atmosphere by the introduction of pollutants from human and natural sources is _______________.

2. Examples of primary air pollutants in nature are dust, sea salt, and _______________.

3. How are secondary pollutants formed?

____

____

____

4. List two examples of secondary pollutants.

____

____

5. What is one reason that ozone near Earth's surface is dangerous?

____

____

6. How is smog formed?

____

____

____

____

7. What is one way local geography plays a part in smog formation in Los Angeles?

____

____

____

HUMAN-CAUSED AIR POLLUTION

______ **8.** How much of the human-caused air pollution in the United States is caused by cars?
a. about 10% to 20%
b. about 25% to 35%
c. about 45% to 55%
d. about 60% to 70%

9. List two sources of industrial air pollution.

____

____

____

10. What are two ways to reduce indoor air pollution?

____

____

ACID PRECIPITATION

______ **11.** Acid precipitation includes rain, sleet, or snow with a high concentration of acids from
a. water pollution.
b. sulfuric acid.
c. air pollution.
d. nitric acid.

______ **12.** When sulfur dioxide and nitrogen oxide combine with water in the atmosphere, they form
a. sulfuric acid and carbon dioxide.
b. sulfuric acid and nitric acid.
c. nitric acid and carbon dioxide.
d. nitric acid and citric acid.

______ **13.** When acid precipitation causes the acidity of soil to increase, it is called
a. calcification.
b. acidification.
c. percolation.
d. deforestation.

14. Where are three of the forest areas in the world that are affected by acid precipitation located?

____

____

____

15. A rapid change in the acidity of a body of water is called

_______________.

THE OZONE HOLE

16. What is the main problem caused by the ozone hole?

____

____

____

17. CFC molecules can remain active in the stratosphere for

_______________.

18. What is one reason the ozone hole is dangerous to humans?

____

____

AIR POLLUTION AND HUMAN HEALTH

19. What are three effects of air pollution on human health?

____

____

____

CLEANING UP AIR POLLUTION

_____ **20.** In the United States, the law that gives the Environmental Protection Agency (EPA) the authority to control the amount of air pollution is the

a. Clean Environment Act.
b. Cleaner World Act.
c. Clean Air Act.
d. Air Quality Act.

21. What are two methods industries use to reduce air pollution?

____

____

22. How is the Allowance Trading System used to reduce air pollution?

____

____

____

23. What are two ways car manufacturers are reducing air pollution?

____

____

24. What are two ways people can reduce pollution from vehicles?

____

____

Name ______________________ Class ______________ Date ______________

Skills Worksheet

Vocabulary and Section Summary B

Characteristics of the Atmosphere

VOCABULARY

After you finish reading the section, try this puzzle! Use the clues below to solve the crossword puzzle.

ACROSS

2. the coldest layer of the atmosphere

5. the layer of the atmosphere where we live

6. the uppermost atmospheric layer

DOWN

1. mixture of gases that surrounds Earth

3. atmospheric layer above the troposphere

4. the measure of the force with which air molecules are pushing on Earth's surface

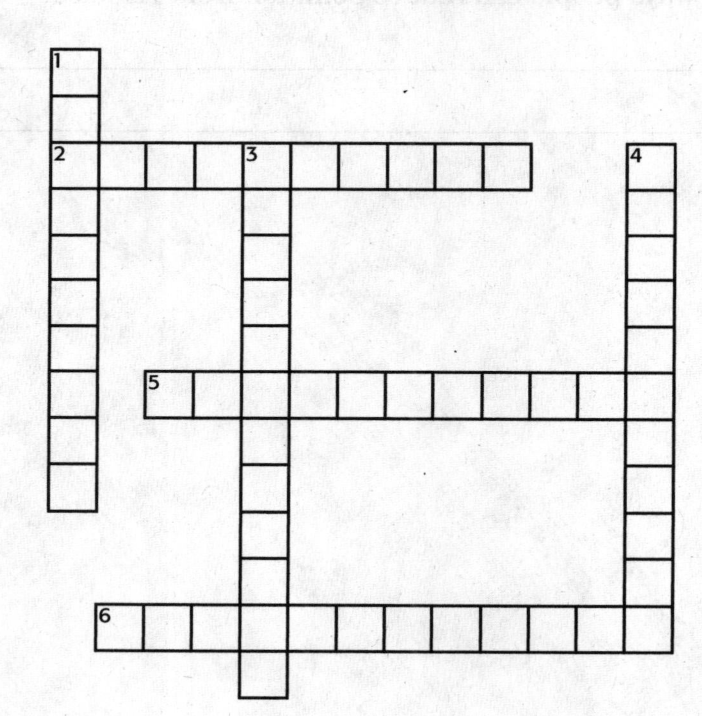

SECTION SUMMARY

Read the following section summary.

- Nitrogen and oxygen make up most of Earth's atmosphere.
- Air pressure decreases as altitude increases.
- The composition of atmospheric layers affects their temperature.
- The troposphere is the lowest atmospheric layer. It is the layer in which we live.
- The stratosphere contains the ozone layer, which protects us from harmful ultraviolet radiation.
- The mesosphere is the coldest atmospheric layer.
- The thermosphere is the uppermost layer of the atmosphere.

Name ______________________ Class ______________ Date ____________

Skills Worksheet

Vocabulary and Section Summary B

Atmospheric Heating

VOCABULARY

After you finish reading the section, try this puzzle! The underlined words are missing all their vowels. Write the completed words in the spaces provided.

1. the transfer of heat due to the movement of a liquid or gas: CNVCTN

   ____

2. all the frequencies or wavelengths of electromagnetic radiation: LCTRMGNTC SPCTRM

   ____

3. the transfer of energy as heat through a material: CNDCTN

   ____

4. the transfer of energy as electromagnetic waves: RDTN

   ____

5. a circular movement of matter that results from differences in density: CNVCTN CRRNT

   ____

6. the warming of Earth's surface and lower atmosphere that occurs when water vapor, carbon dioxide, and other gases absorb and reradiate thermal energy: GRNHS FFCT

   ____

SECTION SUMMARY

Read the following section summary.

- Energy travels from the sun to Earth by radiation. This energy drives many processes at Earth's surface.
- Energy in Earth's atmosphere is transferred by radiation, conduction, and convection.
- Radiation is the transfer of energy through space or matter by waves.
- Conduction is the transfer of energy by direct contact.
- Convection is energy transfer by the movement of matter.

Name ______________________ Class ______________ Date ______________

Skills Worksheet

Vocabulary and Section Summary B

Air Movement and Wind

VOCABULARY

After you finish reading the section, try this puzzle! In the space provided, write the term described. Then, find the words in the word search puzzle on the following page. Words are hidden vertically, horizontally, diagonally, and backward.

______________________ **1.** the movement of air caused by differences in air pressure

______________________ **2.** the curving of the path of a moving object from an otherwise straight path due to Earth's rotation

______________________ **3.** a mixture of gases that surrounds a planet or moon

______________________ **4.** large air-circulation patterns produced by a combination of convection cells, pressure belts, winds, and the Coriolis effect

______________________ **5.** small air-circulation patterns that move short distances and can blow from any direction

______________________ **6.** the transfer of energy as heat through a material

______________________ **7.** the movement of matter due to differences in density and the transfer of energy that results from this movement

______________________ **8.** the transfer of energy as electromagnetic waves

Q	T	F	H	Y	Q	K	M	I	D	J	L	B	B	H	Y	D	L	C	L
O	A	J	Q	U	H	A	M	X	F	U	Y	Q	V	Q	T	J	O	N	O
I	N	U	I	W	Y	I	G	M	J	R	V	M	N	D	C	R	U	N	C
L	Z	T	K	Z	K	N	H	D	B	T	J	T	S	J	I	Z	O	U	A
H	Z	J	H	Z	H	M	E	U	R	D	V	D	F	O	J	I	D	U	L
R	T	U	H	Y	R	V	D	V	Z	E	I	R	L	R	T	P	K	W	W
R	X	S	T	H	S	N	P	U	H	Z	R	I	X	C	W	D	R	T	I
G	L	O	B	A	L	W	I	N	D	S	S	E	U	D	G	I	H	K	N
Z	H	R	B	W	H	S	Y	Q	E	E	V	D	H	M	K	T	N	R	D
K	V	H	W	X	D	L	P	E	F	A	N	R	I	P	R	V	K	D	S
B	M	N	Y	V	X	Q	W	E	C	O	P	I	V	Y	S	O	J	G	I
S	Y	I	U	O	W	O	C	G	C	S	B	B	Z	M	N	O	J	F	L
B	B	M	E	Q	E	T	R	Y	I	G	B	F	N	U	K	O	M	J	V
X	R	Q	R	V	S	P	Y	T	V	M	Q	B	F	V	O	K	U	T	K
R	A	D	I	A	T	I	O	N	G	S	Y	X	S	G	W	T	R	A	A
C	B	U	E	W	T	Z	X	R	L	Z	B	C	W	V	M	H	H	A	R
N	E	M	A	N	F	P	L	L	J	Q	B	O	M	L	M	W	W	H	A
J	K	Z	G	Z	W	H	Y	L	C	P	Z	S	J	T	V	G	R	H	C
E	H	S	F	S	Y	F	U	N	O	I	T	C	E	V	N	O	C	D	L

SECTION SUMMARY

Read the following section summary.

- Winds blow from areas of high pressure to areas of low pressure.
- Pressure belts are caused by the uneven heating of Earth's surface by the sun.
- The Coriolis effect causes wind to appear to curve as it moves across Earth's surface.
- Global winds include the polar easterlies, the westerlies, and the trade winds.
- Local winds include sea and land breezes and valley and mountain breezes.

Name ______________________ Class ______________ Date ______________

Skills Worksheet

Vocabulary and Section Summary B

The Air We Breathe

VOCABULARY

After you finish reading the section, try this puzzle! Look at the clues below, and write the terms being described in the blanks provided. Then, write the boxed letters in the space provided to spell out a phrase related to the atmosphere.

1. a chemical that causes ozone to break down into oxygen, which does not block harmful UV rays

2. pollutants that are put directly into the air by human or natural activity

3. a device that is used to remove some pollutants before they are released by smokestacks

4. the contamination of the atmosphere by the introduction of pollutants from human and natural sources

5. rain, sleet, or snow that contains a high concentration of acids

6. the mixture of indoor air with outdoor air

7. the increase of soil acidity due to acid precipitation

8. pollutants that form when primary pollutants react with other primary pollutants or with naturally occurring substances, such as water vapor

9. a rapid change in the acidity of a body of water

1. _ _ ☐

2. _ _ _ _ _ _ _ _ _ ☐ _ _ _ _ _ _ _ _

3. _ _ _ _ _ _ _ _ ☐ _

4. ☐ _ _ _ _ _ _ _ _ _ _ _ _ _

5. _ _ _ _ _ _ _ _ _ _ _ _ _ _ _ _ ☐

6. _ ☐ _ _ _ _ _ _ _ _ _

7. ☐ _ _ _ _ _ _ _ _ _ _ _ _

8. _ _ _ _ _ _ _ _ _ ☐ _ _ _ _ _ _ _ _ _ ☐ _

9. _ _ _ _ ☐ _ _ _

10. What is the phrase?

____

SECTION SUMMARY

Read the following section summary.

- Air pollution is the introduction of harmful substances into the air by humans or by natural events.
- Primary pollutants are pollutants that are put directly into the air by human or natural activity.
- Secondary pollutants are pollutants that form when primary pollutants react with other primary pollutants or with naturally occurring substances.
- Transportation, industry, and natural sources are the main sources of air pollution.
- The burning of fossil fuels may lead to air pollution and acid precipitation, which may harm human and wildlife habitats.
- Air pollution can be reduced by legislation, such as the Clean Air Act; by technology, such as scrubbers; and by changes in lifestyle.

Name ______________________ Class ______________ Date ____________

Skills Worksheet

Directed Reading B

Section: Water in the Air (pp. 506–511)

1. The condition of the atmosphere at a certain time and place is called

______________________.

THE WATER CYCLE

2. What heats Earth's surface and causes water to change states?

____

3. What is the water cycle?

____

____

____

HUMIDITY

Match the correct definition with the correct term. Write the letter in the space provided.

_____ **4.** the temperature at which the rate of condensation equals the rate of evaporation

_____ **5.** the ratio of the amount of water vapor in the air to the amount of water vapor needed to reach saturation

_____ **6.** that part of total atmospheric pressure that is caused by water vapor

_____ **7.** the amount of water vapor in the air

a. humidity
b. relative humidity
c. vapor pressure
d. dew point

8. What does humidity depend on?

____

____

9. What happens when the rate of evaporation equals the rate of condensation?

____

____

10. When temperatures are below the dew point, what forms on a surface?

____

11. Explain how using a humidity sensor with a thin-film polymer measures humidity.

____

____

____

CONDENSATION

______ **12.** The process by which a gas becomes a liquid is called

a. humidity.
b. condensation.
c. water vapor.
d. saturation.

13. When is air saturated?

____

____

14. Why do water droplets form on the outside of a glass of ice water?

____

____

____

15. What is dew?

____

____

CLOUDS AND PRECIPITATION

16. What is a cloud made of?

____

____

17. What are two ways in which clouds are classified?

____

____

18. Name the three cloud forms.

____

____

____

19. Water, in any form, that falls to Earth's surface from the clouds is called

______________________.

20. Name the four major forms of precipitation.

____

____

____

____

Name _______________ Class _______________ Date _______________

Skills Worksheet

Directed Reading B

Section: Fronts and Weather (pp. 512–519)

1. What causes changes in the weather?

____

____

2. A large body of air that has the same temperature and moisture content throughout is called a(n) _______________.

FRONTS

_____ **3.** What usually happens when different air masses meet?
- **a.** Cold air rises.
- **b.** Warm air rises.
- **c.** The masses disappear.
- **d.** Air from the two masses mixes together.

_____ **4.** The boundary between air masses of different densities is called a(n)
- **a.** surface current.
- **b.** cloud.
- **c.** front.
- **d.** anticyclone.

Match the correct description with the correct term. Write the letter in the space provided.

_____ **5.** A warm air mass moves over a cold, denser air mass.

_____ **6.** A warm air mass is caught between two colder air masses.

_____ **7.** A cold air mass meets a warm air mass, but the two remain separated.

_____ **8.** A cold air mass moves under a warm, less dense air mass.

a. cold front

b. warm front

c. occluded front

d. stationary front

9. Describe the typical weather brought by each front listed below.

Cold front: ______________________________

____

Warm front: ______________________________

____

Occluded front: ______________________________

____

Stationary front: ______________________________

____

AIR PRESSURE AND WEATHER

10. An area in the atmosphere that has lower air pressure than the surrounding areas and has winds that spiral toward the center is called

a(n) ____________________.

11. Areas of high air pressure are called ____________________.

12. How does a cyclone affect the weather?

____

____

13. How does an anticyclone affect the weather?

____

____

THUNDERSTORMS

______ **14.** What is an intense local storm that forms strong winds, heavy rain, lightning, and thunder called?

a. anticyclone
b. storm surge
c. tornado
d. thunderstorm

15. What two atmospheric conditions produce thunderstorms?

____

____

16. An electric discharge between two oppositely charged surfaces is

______________________.

17. The sound caused by air rapidly expanding along a lightning strike is

called ______________________.

TORNADOES

_____ **18.** What is a rapidly spinning air column with very high wind speeds that touches the ground?

a. thunderstorm
b. tornado
c. anticyclone
d. occluded front

19. When does a funnel cloud become a tornado?

____

____

HURRICANES

20. A large, rotating tropical weather system with wind speeds of at least

120 km/h is called a(n) ______________________.

21. Where do most hurricanes form?

____

____

____

22. Where does a hurricane get its energy?

____

____

23. Why doesn't California experience hurricanes?

____

____

____

____

Name ______________________ Class ______________ Date ____________

EFFECTS OF SEVERE WEATHER

24. Name five things that are part of severe weather.

____

____

____

____

____

25. What is the leading cause of weather-related deaths?

____

26. A rise in sea level that forms in the ocean during a storm is called

a(n) ______________________.

SEVERE-WEATHER SAFETY

27. What is the most important safety measure during a flood warning?

____

____

Name ______________________ Class ______________ Date ____________

Skills Worksheet

Directed Reading B

Section: What Is Climate? (pp. 520–527)

CLIMATE VERSUS WEATHER

______ **1.** Weather conditions
- **a.** change from day to day.
- **b.** never change.
- **c.** change slowly over a long period of time.
- **d.** change every two weeks.

2. The average weather conditions in an area over a long period of time is

______________________.

3. What are the two main factors that determine climate?

____

____

FACTORS THAT AFFECT CLIMATE

4. Name the four factors that affect the temperature and precipitation rates of an area.

____

____

____

____

5. What does latitude measure?

____

____

6. Explain how latitude affects the temperatures at the poles.

____

____

7. What affects the amount of direct solar energy at different latitudes?

____

____

8. Explain why seasons happen.

____

____

____

9. What are the patterns of winds the result of?

____

____

10. How do winds affect climate and weather?

____

____

____

11. The sizes and shapes of the land-surface features of a region combine to form ______________________.

12. At high ______________________, the temperature of the air is lower.

13. Explain the process that causes a rain-shadow effect.

____

____

____

14. How does the ocean keep temperatures in California moderate?

____

____

____

15. Streamlike movements of water at or near the surface of the ocean are called ______________________.

16. What carries cold water from the northern Pacific Ocean south to Mexico?

____

CLIMATES OF THE WORLD

17. Why do polar bears live only in very cold arctic regions?

____

____

18. Name Earth's three major climate zones.

____

____

____

19. Why are there several kinds of climates within each of the three major climate zones?

____

____

____

Name ______________________ Class ______________ Date ____________

Skills Worksheet

Directed Reading B

Section: Changes in Climate (pp. 528–533)

1. Why would you not notice a change in climate?

____

____

2. What two factors can make the climate change?

____

____

ICE AGES

3. What does the geologic record show about Earth's climate?

____

____

4. A long period when ice forms in high latitudes and moves toward lower latitudes is a(n) ____________________.

5. Periods of cold and periods of warmth during an ice age are called ____________________.

6. Why does the sea level drop during times when ice covers large areas of Earth?

____

____

CAUSES OF CLIMATE CHANGE

7. What does the Milankovitch theory explain?

____

____

____

8. What does a continent's location relative to the equator and poles determine?

____

____

9. What happens when continents move?

____

____

10. How does the sun's energy affect the amount of energy in Earth's system?

____

____

____

____

11. A small, rocky object that orbits the sun is called a(n)

______________________.

12. Dust and smaller rocks are called ______________________.

13. Explain how an asteroid could have caused the dinosaurs to become extinct 65 million years ago.

____

____

____

____

____

14. What happens during a volcanic eruption in which some of the sun's rays are blocked?

____

____

15. A gradual increase in average global temperatures is called

______________________.

GLOBAL WARMING

______ **16.** Earth's natural heating process, in which gases in the atmosphere absorb thermal energy, is called

- **a.** global warming.
- **b.** the greenhouse effect.
- **c.** an ice age.
- **d.** a glacial period.

17. What do many scientists think cause the rise in global temperatures?

____

____

18. List three possible consequences of global warming.

____

____

____

Name ____________ Class ____________ Date ____________

Skills Worksheet

Vocabulary and Section Summary B

Water in the Air

VOCABULARY

After you finish reading the section, try this puzzle! In each of the following items, use the clue to unscramble the letters, and write the term in the space provided.

1. amount of moisture the air contains compared with the maximum it can hold at a given temperature: TIVEALER YTIDIMUH

____

2. water in solid or liquid form that falls from the air to Earth: CIPIPERATTOIN

____

3. amount of moisture in the air: DIMUHTIY

____

4. condition of the atmosphere at a certain time and place: EAWETHR

____

5. when water vapor becomes a liquid: OEAIOCNDNSTN

____

6. temperature at which the rate of condensation equals the rate of evaporation: EDW OIPTN

____

SECTION SUMMARY

Read the following section summary.

- The sun's energy causes water to change states and to move through the water cycle.
- The amount of water vapor in the air is called *humidity*.
- When the temperature of the air cools to the dew point, the air is saturated and liquid water droplets form.
- Clouds form as air rises and cools, which causes water droplets to form on small particles in the air.
- Precipitation is water, in any form, that falls to Earth's surface from the clouds.

Name ______________________ Class ______________ Date ____________

Skills Worksheet

Vocabulary and Section Summary B

Fronts and Weather

VOCABULARY

After you finish reading the section, try this worksheet!

Donald Deluge is a storm chaser—a person who follows dangerous weather in order to observe and record its effects. He likes to report his observations over the radio from his storm-chaser van. Unfortunately, some of the words from his latest broadcast were hard to understand, and Don hasn't been seen since his last adventure, so he isn't around to clear things up. See if you can help by reading the transcript of Don's latest adventure and filling in the blank spaces with the vocabulary words listed below.

Vocabulary Words

- tornado
- front
- thunder
- cyclone
- air mass
- lightning

[begin transcript]
12:34 pm

It seems there is a large ____________________ to the north of our current position in which the warm temperature and high humidity of the air are constant throughout. From the south, there is a cold ____________________ moving under this warmer area. Looking at the satellite images, there appears to be a low pressure zone in which air is spiraling in and forming a(n) ____________________. There's a lot of cloud cover forming there. Looks like we might get a nice-sized storm! I'll be in contact again once I get closer to the hot zone.

1:21 pm

I was right! We have a nice thunderstorm forming. Looks like a cold front has moved in over the mountains to the west, right on top of the warm mass. There's a lot of electrical energy in the air here. The cumulonimbus clouds must have an opposite charge than the ground surface because all kinds of ____________________ strikes are occurring around me. . . WOW! Did you hear that ____________________? A tree right next to us just got hit! This storm is awesome! I'll be back in contact once we clear the tree from the road.

1:38 pm

A funnel cloud is forming above a field to the right of us. Looks like it is going

to touch down and we might have ourselves a(n) ____________________!

We're going to try to get a little closer. I'll be in contact again shortly.

[end transcript]

SECTION SUMMARY

Read the following section summary.

- Thunderstorms are weather systems that produce strong winds, heavy rain, lightning, and thunder.
- Lightning is a large electric discharge that occurs between two oppositely charged surfaces. Lightning releases a great deal of energy and can be very dangerous.
- Tornadoes are small, rapidly rotating columns of air that touch the ground and can cause severe damage.
- A hurricane is a large, rotating tropical weather system that has high wind speeds.
- In the event of severe weather, it is important to stay safe. Listen to your local TV or radio stations for updates, and remain indoors and away from windows.

Name ______________________ Class ______________ Date ______________

Skills Worksheet

Vocabulary and Section Summary B

What Is Climate?

VOCABULARY

After you finish reading the section, try this puzzle! In the space provided, write the term described. Use the clues below to fill in the correct term, then solve the word search puzzle below. Words in the puzzle are hidden vertically, horizontally, diagonally, and backward.

______________ **1.** the average weather condition in an area over a long period of time

______________ **2.** the condition of the atmosphere at a certain time

______________ **3.** the distance north or south from the equator

______________ **4.** the height of surface landforms above sea level

______________ **5.** a horizontal movement of ocean water that occurs at or near the ocean's surface

______________ **6.** the combination of the sizes and shapes of the land-surface features of a region

S	U	R	F	A	C	E	C	U	R	R	E	N	T
A	V	W	B	S	T	B	L	U	A	E	I	O	O
P	E	W	T	W	S	M	I	N	F	G	M	A	P
C	L	I	S	P	E	A	M	X	E	I	B	T	O
P	E	H	M	N	A	A	A	A	M	L	K	G	G
H	V	B	A	R	O	H	T	Y	E	B	B	L	R
S	A	C	B	I	O	N	E	H	E	A	C	M	A
F	T	T	C	A	H	D	I	D	E	I	O	G	P
H	I	S	L	E	G	L	M	O	S	R	E	R	H
P	O	N	B	A	F	E	S	M	B	R	A	S	Y
N	N	D	M	E	D	U	T	I	T	A	L	F	G

SECTION SUMMARY

Read the following section summary.

- Climate is the average weather conditions in an area over a long period of time.
- The curve of Earth's surface affects the amount of direct solar energy that reaches the ground at different latitudes.
- The tilt of Earth on its axis causes the seasons.
- Winds affect the climate of an area by transferring solar energy and by carrying moisture.
- Topography influences an area's climate by affecting both the transfer of energy and the rate of precipitation.
- Large bodies of water and ocean currents influence the climate of an area by affecting the temperature of the air.
- The three main climate zones of the world are the tropical zone, the temperate zone, and the polar zone.

Name ______________________ Class ______________ Date ______________

Skills Worksheet

Vocabulary and Section Summary B

Changes in Climate

VOCABULARY

After you finish reading the section, try this puzzle! The terms in parentheses have all lost their vowels and spaces. Use the clues to fill in the correct term in the space provided.

1. During ______________________, large sheets of ice advance, and the sea level drops. (GLCLPRDS)

2. Natural events and human activities can cause ______________________, which is a gradual increase in the average global temperature. (GLBLWRMNG)

3. When a large piece of rock hits Earth, ______________________ shoots into the atmosphere. (DBRS)

4. Earth's natural heating process is called the ______________________. (GRNHSFFCT)

5. Scientists have found evidence of many major ______________________ during Earth's geologic history. (CGS)

6. A small, rocky object that orbits the sun and sometimes crashes into Earth is called a(n) ______________________. (STRD)

SECTION SUMMARY

Read the following section summary.

- Earth's climate has changed repeatedly throughout geologic time.
- The Milankovitch theory states that Earth's climate changes as Earth's orbit and the tilt of Earth's axis change.
- Climate changes may also be affected by continental drift, volcanic eruptions, asteroid impacts, solar activity, and human activity.
- Excess CO_2 is believed to contribute to global warming.

Name ________________ Class ________________ Date ________________

Skills Worksheet

Directed Reading B

Section: Everything Is Connected (pp. 550–553)

1. What is one way in which two organisms interact?

____

STUDYING THE WEB OF LIFE

2. What is ecology?

____

____

3. What is the abiotic part of the environment?

____

____

4. What is the biotic part of the environment?

____

For each word listed, write whether it is a biotic or abiotic part of the environment.

________________ 5. black bear

________________ 6. fish

________________ 7. temperature

________________ 8. water

________________ 9. plants

________________ 10. rocks

11. Name the six levels of the environment.

____

____

____

____

____

____

12. A group of turtles competing for food, shelter, and mates in a salt marsh is an example of a(n) ______________________.

13. What is a community?

____

____

____

____

____

14. An ecosystem is made up of a community of organisms and the ______________________ parts of the environment, such as temperature, soil, and water.

15. In a biome, what determines the kinds of plants and animals that live there?

____

16. The ocean, the air, and all areas of Earth where life exists are all parts of the ______________________.

Name ______________________ Class ______________ Date ____________

Skills Worksheet

Directed Reading B

Section: Living Things Need Energy (pp. 554–559)

______ **1.** To survive, living things need
- **a.** grasslands.
- **b.** energy.
- **c.** clothing.
- **d.** species.

THE ENERGY CONNECTION

2. What three groups can animals be divided into based on how they get energy?

____

____

____

3. Organisms that change the energy in sunlight into food are called

_______________________.

4. Producers use a process called _______________________ to make food.

5. Herbivores, carnivores, and omnivores are known

as _______________________.

6. A consumer that eats only plants is a(n) _______________________.

7. A consumer that eats only animals is a(n) _______________________.

8. A consumer that eats both plants and animals is

a(n) _______________________.

9. Omnivores that eat dead plants and animals are called

_______________________.

10. Organisms that break down dead organisms to get energy are called

_______________________.

11. List two decomposers.

____

____

12. A diagram that shows how energy in food is transferred from one organism

to another is a(n) _______________________.

13. Name the three types of consumers.

____

____

____

14. What is a food web?

____

____

15. What happens to the energy not immediately used by an organism?

____

____

____

16. The diagram that shows an ecosystem's loss of energy at each level of the food chain is known as a(n) ______________________.

17. Explain why less energy is available at higher levels of an energy pyramid.

____

____

WOLVES AND THE ENERGY PYRAMID

18. When gray wolves were almost wiped out as the wilderness was settled, what happened to the grass and elk in some areas?

____

____

____

____

19. What happened when gray wolves were brought back to Yellowstone National Park?

____

____

____

____

20. How are all organisms in a food web important?

____

____

____

21. Why are ranchers near Yellowstone worried about wolves returning?

____

____

Name ______________________ Class ______________ Date ______________

Skills Worksheet

Directed Reading B

Section: Types of Interactions (pp. 560–565)

______ **1.** In a natural community, population sizes vary because
- **a.** the populations are not affected by each other.
- **b.** the populations all affect one another.
- **c.** individuals in the populations decide to have big families.
- **d.** the populations are able to grow without stopping.

INTERACTIONS WITH THE ENVIRONMENT

2. When a frog lays hundreds of eggs in a small pond, what happens to the population of frogs in the pond? Explain your answer.

____

____

3. A resource so scarce that it limits the size of a population is called a(n) ______________________.

4. In what way can food become a limiting factor?

____

____

____

5. The largest population that an environment can support is called its ______________________.

6. When a population grows larger than its carrying capacity, what happens?

____

____

____

INTERACTIONS BETWEEN ORGANISMS

7. Ecologists have listed three main ways that species and individuals affect each other—competitive relationships, prey and ______________________ relationships, and symbiotic relationships.

COMPETITION

8. When two or more individuals or populations try to use the same resource, it is called ______________________.

9. Give one example of competition between individuals within a population.

____

10. Give one example of competition between populations.

____

PREDATORS AND PREY

_____ **11.** What is an organism that kills and eats another organism called?
- **a.** predator
- **b.** carrier
- **c.** competitor
- **d.** prey

_____ **12.** What is an organism that is eaten by another organism called?
- **a.** predator
- **b.** carrier
- **c.** competitor
- **d.** prey

13. List two adaptations predators use to catch prey.

____

____

14. List two ways prey have adapted to avoid predators.

____

____

15. What is blending in with the background called?

____

16. Give two examples of animals using defensive chemicals against predators.

____

____

17. How can being bright red, yellow, or orange help an animal avoid predators?

____

____

SYMBIOSIS

_____ **18.** A close, long-term association between two or more species is called
- **a.** symbiosis.
- **b.** defensive chemicals.
- **c.** predator adaptations.
- **d.** camouflage.

19. What are the three types of symbiosis?

____

____

____

20. Both organisms benefit in the type of symbiosis called

_______________________.

21. When one organism benefits and the other is unaffected, the symbiotic relationship is called _______________________.

22. A symbiotic relationship in which one organism benefits and the other is harmed is called _______________________.

23. In parasitism, the organism that benefits is called the _______________________.

24. The organism that is harmed by a parasite is called the _______________________.

Name ________________________ Class ________________ Date ____________

Skills Worksheet

Vocabulary and Section Summary B

Everything Is Connected

VOCABULARY

After you finish reading the section, try this puzzle! Use the clues below to figure out what vocabulary terms go in each space of the puzzle below.

1. the study of the interactions between organisms and their environment

___ ___ ___ ___ ___ ___ ___
(7 under the 4th space)

2. a group of individuals of the same species that live together in the same area at the same time

___ ___ ___ ___ ___ ___ ___ ___ ___ ___
(2 under the 10th space)

3. nonliving factors in the environment

___ ___ ___ ___ ___ ___ ___
(6 under the 1st space)

4. living factors in the environment

___ ___ ___ ___ ___ ___
(1 under the 2nd space)

5. the part of Earth where life exists

___ ___ ___ ___ ___ ___ ___ ___ ___
(4 under the 2nd space)

6. several populations living in the same area

___ ___ ___ ___ ___ ___ ___ ___ ___
(5 under the 5th space)

7. a community of organisms and their nonliving environment

___ ___ ___ ___ ___ ___ ___ ___ ___
(8 under the 4th space)

8. an area in which the climate determines the plant community

___ ___ ___ ___ ___
(3 under the 2nd space)

Write each of the numbered letters in the correct space below to find the first level of environmental organization.

___	___	D	___	V	___	D	___	___	___	___
1	2		3		4		5	6	7	8

SECTION SUMMARY

Read the following section summary.

- Organisms in an ecosystem depend on other organisms and on abiotic factors for their survival.
- Energy and other resources flow between organisms and their environment.
- Biotic factors are the interactions between organisms in an area, such as competition.
- Abiotic factors include all of the nonliving things in an area, such as water and light.

Name ______________________________ Class ________________ Date ______________

Skills Worksheet

Vocabulary and Section Summary B

Living Things Need Energy

VOCABULARY

After you finish reading the section, try this puzzle! Use the clues below to solve the crossword puzzle. Then, unscramble the highlighted letters to answer the bonus question.

ACROSS

5. shows the loss of energy in an ecosystem

DOWN

1. shows the pathway of energy transfer between a series of organisms

2. shows the feeding relationships in an ecosystem

3. eat only plants

4. eat only animals

BONUS QUESTION

What type of consumer are most human beings?

___ ___ ___ ___ ___ ___ ___ ___ ___

SECTION SUMMARY

Read the following section summary.

- Producers that use photosynthesis transfer the energy in sunlight into chemical energy.
- Consumers eat producers and other organisms to obtain energy and nutrients.
- Decomposers break down all of the materials in dead organisms to obtain energy and nutrients.
- Food chains represent how energy and nutrients are transferred from one organism to another.
- Energy pyramids show how energy is lost at each level of the food chain.
- All organisms have important roles in a food web.

Name ______________________ Class ______________ Date ____________

Skills Worksheet

Vocabulary and Section Summary B

Types of Interactions

VOCABULARY

After you finish reading the section, try this puzzle! In the space provided, write the term described. Then, find the words in the puzzle on the next page. Words may be hidden vertically, horizontally, diagonally, and backward.

______________ **1.** an organism that is killed and eaten by another organism

______________ **2.** an organism that kills and eats another organism

______________ **3.** the largest population that an environment can support

______________ **4.** relationship in which one organism benefits and the other is unaffected

______________ **5.** relationship in which two different organisms live in close association

______________ **6.** relationship in which both species benefit

______________ **7.** relationship in which one species benefits and the other is harmed

SECTION SUMMARY

Read the following section summary.

- Limiting factors refer to the factors in the environment that keep a population from growing without limits.
- Competition happens within populations and between populations.
- A predator is an organism that kills an organism to eat all or part of that organism. The organism that is killed and eaten is called *prey*.
- Prey have features such as camouflage, chemical defenses, and warning coloration that protect them from predators.
- Mutualism, commensalism, and parasitism are the three kinds of symbiotic relationships that occur between organisms.

Vocabulary and Section Summary B *continued*

Y	G	W	M	O	U	M	H	F	M	P	X	M	B	A	A
V	T	M	D	Q	J	D	Z	S	C	R	G	S	C	M	T
D	B	I	X	S	N	D	I	E	L	E	Y	I	F	D	M
Q	F	E	C	M	F	T	X	H	V	D	P	L	X	Z	V
M	U	X	P	A	I	U	T	O	F	A	D	A	D	K	T
M	L	R	P	S	P	Z	M	R	T	T	M	S	L	N	E
O	E	P	A	H	M	A	Q	C	F	O	S	N	K	U	M
Y	K	R	H	K	Y	U	C	G	G	R	O	E	Y	E	W
M	A	S	H	Q	I	B	T	G	Y	W	J	M	E	K	H
P	S	Q	R	R	G	X	G	U	N	Y	U	M	U	R	D
S	Y	M	B	I	O	S	I	S	A	I	I	O	E	U	A
U	I	V	W	Z	B	A	D	N	J	L	Y	C	H	Z	N
H	J	M	W	V	X	L	N	O	G	J	I	R	G	W	T
Q	K	C	L	W	F	J	O	T	Y	D	T	S	R	A	W
O	G	M	S	L	L	Q	J	I	N	I	Q	U	M	A	R
R	N	Y	G	G	L	K	E	P	C	I	Y	Z	E	Q	C

Name ______________________ Class ______________ Date ____________

Skills Worksheet

Directed Reading B

Section: Studying the Environment (pp. 580–583)

1. A nonliving part of the environment is a(n) ______________________ factor.

2. What is a living part of the environment called?

____

HOW BIOMES DIFFER FROM ECOSYSTEMS

______ **3.** Biomes are made up of many connected
- **a.** deposits.
- **b.** organisms.
- **c.** ecosystems.
- **d.** communities.

______ **4.** What defines an ecosystem?
- **a.** plant communities
- **b.** water flow
- **c.** organisms and abiotic factors
- **d.** land boundaries

5. The kinds of plants in a biome are determined by the

______________________ of the biome.

6. Name two abiotic factors that determine climate.

____

____

ABIOTIC FACTORS OF AN ENVIRONMENT

Match the correct description with the correct term. Write the letter in the space provided.

______ **7.** has nutrients that help plants grow

______ **8.** provides energy that is used by plants to produce food

______ **9.** determines what types of animals live in an area

______ **10.** determines the amount of plant growth in an area

- **a.** sunlight
- **b.** temperature
- **c.** rainfall
- **d.** soil

ROLES OF ORGANISMS IN AN ENVIRONMENT

______ **11.** Organisms that convert energy in the environment into food are called
a. producers.
b. abiotic factors.
c. decomposers.
d. consumers.

______ **12.** Organisms that feed on other organisms are called
a. producers.
b. energy factors.
c. herbivores.
d. consumers.

______ **13.** Organisms that break down dead organisms are called
a. plants.
b. abiotic factors.
c. decomposers.
d. consumers.

14. Why might decomposition take thousands of years in Antarctica?

____

____

SIMILAR CLIMATES, SIMILAR BIOMES

15. Why do some separated biomes have similar plants and animals?

____

16. List four factors that can cause variations in climate.

____

____

____

17. What is the biome found in much of California called?

____

18. A biome that is similar to the biome found in most of California is located in which part of the world?

____

____

Name ______________________ Class ______________ Date ____________

Skills Worksheet

Directed Reading B

Section: Land Biomes (pp. 584–591)

1. What are the three categories into which each organism in a biome can be placed?

____

____

____

DESERTS

_____ **2.** A region that is very dry and very hot is called a
- **a.** prairie.
- **b.** grassland
- **c.** desert.
- **d.** savanna.

3. List two adaptations of desert plants.

____

____

Match the correct description with the correct term. Write the letter in the space provided.

_____ **4.** stores water under its shell

_____ **5.** buries itself in loose sand

a. fringe-toed lizard

b. desert tortoise

CHAPARRAL

6. What are the summers and winters like in the chaparral?

____

____

7. Plants that keep their leaves all year round are ______________ plants.

8. What abiotic factor helps maintain the chaparral?

____

____

9. How do chipmunks and mule deer adapt to the chaparral biome?

____

____

GRASSLANDS

______ **10.** Which of the following describes a grassland?

a. grasses and small plants
b. permafrost
c. forest canopy
d. evergreen shrubs

11. Name three things that prevent the growth of trees and shrubs in temperate grasslands.

____

____

____

12. A grassland with clumps of trees and seasonal rains is called a(n) ______________________.

13. Give an example of an animal that preys on herbivores in a savanna.

____

TUNDRA

14. A biome with very cold temperatures and little rainfall is called a(n) ______________________.

15. The layer of soil in a polar tundra that is frozen all year round is called ______________________.

16. Why are shallow-rooted plants, such as grasses and small shrubs, common in a polar tundra?

____

____

17. In the tundra, mosquitoes lay eggs in the mud during summer. What animals prey on these insects?

____

FORESTS

18. Name the three main types of forest biomes.

____

____

____

19. Most of the trees in a coniferous forest are called ______________.

20. What are two traits of conifers?

____

____

21. What are two reasons that few large plants grow beneath the trees of a coniferous forest?

____

____

22. A tree whose leaves fall off is a(n) ______________ tree.

23. The layer of the forest in the tree tops is called the ______________.

24. What does it mean when we say that tropical rain forests are the most diverse places on Earth?

____

____

25. Where do most animals live in the rain forest?

____

26. Name one carnivore of the tropical rain forest.

____

27. Where are most of the nutrients in a tropical rain forest?

____

Name ______________________ Class ______________ Date ____________

Skills Worksheet

Directed Reading B

Section: Marine Ecosystems (pp. 592–599)

1. Ecosystems in the ocean are called ______________________.

DEPTH AND SUNLIGHT

2. What are two abiotic factors that shape marine ecosystems?

____

____

3. How far does sunlight reach in the oceans?

____

4. Tiny organisms that float near the surface of the ocean are called

______________________.

5. What two things form the base of most of the ocean's food chains?

____

____

TEMPERATURE

______ **6.** How does the temperature of ocean water change as it becomes deeper?

a. It becomes colder.
b. It becomes warmer.
c. It does not change.
d. It varies from day to day.

______ **7.** Which ocean temperature zone has the warmest water?

a. deep zone
b. thermocline
c. surface zone
d. middle layer

______ **8.** If the water is too hot or too cold, barnacles

a. can eat.
b. can get large.
c. may migrate.
d. may die.

MAJOR ZONES IN THE OCEAN

Match the correct description with the correct term. Write the letter in the space provided.

______ **9.** warm water and lots of sunlight; ocean floor starts to slope downward

______ **10.** where the sea floor drops sharply; deep waters of the open ocean

______ **11.** ocean floor; no sunlight

______ **12.** where the ocean meets the shore; exposed to air part of the day

a. intertidal zone

b. neritic zone

c. oceanic zone

d. abyssal zone

13. What do consumers in the abyssal zone feed on?

____

____

KINDS OF MARINE ECOSYSTEMS

______ **14.** Most of the water that makes up Earth's precipitation is provided by

a. ponds and lakes.

b. the ocean.

c. estuaries.

d. swamps.

15. Name three kinds of intertidal ecosystems.

____

____

____

16. Seaweed use ______________________ **to attach themselves to rocks.**

17. A place where fresh water from rivers mixes with salt water from the ocean is a(n) ______________________.

18. What must organisms in estuaries be able to survive?

____

19. Where are most coral reefs found?

____

20. The structure of a coral reef is based on the skeletons of ______________________.

21. Name three marine organisms that live on coral reefs.

____

____

____

22. Where are kelp forests found?

____

23. What do sea otters do to stay afloat while they take a nap?

____

24. Why are vent worms unusual?

____

____

25. An ecosystem found only in the middle of the Atlantic Ocean is called the

______________________.

26. What are sargassums?

____

____

27. How do animals in the Sargasso Sea hide from predators?

____

____

28. Why are there large numbers of phytoplankton that are able to support many organisms in the polar ice ecosystem?

____

____

____

Name ______________________ Class ______________ Date ____________

Skills Worksheet

Directed Reading B

Section: Freshwater Ecosystems (pp. 600–603)

1. Name three examples of freshwater ecosystems.

____

____

____

2. What four abiotic factors affect freshwater ecosystems?

____

____

____

____

STREAM AND RIVER ECOSYSTEMS

______ **3.** What is a place where water flows from underground to Earth's surface called?

a. spring
b. swamp
c. tributary
d. river

4. A stream of water that joins a larger stream is called a(n)

______________________.

5. A very strong, wide stream is called a(n) ______________________.

6. How do algae and moss adapt to fast-moving water?

____

____

____

POND AND LAKE ECOSYSTEMS

Match the correct description with the correct term. Write the letter in the space provided.

______ **7.** zone that goes as deep as sunlight can reach; home to many photosynthetic plankton

______ **8.** zone where no sunlight reaches; scavengers feed on dead organisms from above

______ **9.** zone closest to the edge of a lake or pond; sunlight reaches the bottom

a. deep-water zone
b. open-water zone
c. littoral zone

10. What makes it possible for algae and plants to grow in the littoral zone?

____

____

____

11. Name two scavengers that live in the deep-water zone.

____

____

WETLAND ECOSYSTEMS

______ **12.** A wetland is an area of land that
a. has only trees.
b. is very dry.
c. is sometimes under water.
d. is never flooded.

Match the correct definition with the correct term. Write the letter in the space provided.

______ **13.** a wetland ecosystem in which shrubs and trees grow

______ **14.** a treeless wetland ecosystem where plants such as grasses grow

a. marsh
b. swamp

15. Name three reasons that wetlands are valuable.

____

____

____

16. Where are freshwater marshes often found?

____

____

17. Where are swamps found?

____

____

HOW AN ECOSYSTEM CAN CHANGE

_____ **18.** How does a lake start to become a forest?

- **a.** Plants grow farther from the center of the lake.
- **b.** Sediment and leaves settle at the bottom of the lake.
- **c.** The lake becomes a stream.
- **d.** Fishes die off due to increased oxygen levels.

_____ **19.** What happens when decaying plant and animal life at the bottom of a lake decompose?

- **a.** Animal life in the lake increases.
- **b.** The lake has more water.
- **c.** Oxygen is lost, affecting the animals that live in the lake.
- **d.** Living conditions for fish improve.

Name ______________________ Class ______________ Date ____________

Skills Worksheet

Vocabulary and Section Summary B

Studying the Environment

VOCABULARY

After you finish reading the section, try this puzzle! In the space provided, write the term described. Then, unscramble the boxed letters to answer the question about the environment.

1. large area made up of many connected ecosystems
2. all the populations of species that live in an area
3. kind of biome found in much of California and the area around the Mediterranean Sea
4. consumer that eats only plants
5. consumer that eats only animals
6. made up of a combination of the average yearly temperature and the average yearly rainfall
7. consumer that eats both plants and animals

1. ☐ _ ☐ _ _
2. ☐ _ _ _ _ _ _ ☐ _
3. ☐ _ ☐ _ _ _ _ _ _ _
4. _ _ _ _ ☐ _ _ _ _
5. _ ☐ _ _ _ _ ☐ _ _
6. _ _ _ _ _ ☐ _
7. _ _ _ ☐ _ _ ☐ _

8. What is a nonliving part of the environment called?

_ _ _ _ _ _ _ F _ _ _ _ _

SECTION SUMMARY

Read the following section summary.

- Biomes are made up of many connected ecosystems.
- Some abiotic factors are resources. Other abiotic factors are conditions in the environment.
- Organisms in biomes and ecosystems can be categorized as producers, consumers, or decomposers.
- Some widely-separated biomes have similar communities of plants and animals.
- Similar biomes are found in areas that have similar climates.

Name ______________________ Class ______________ Date ______________

Skills Worksheet

Vocabulary and Section Summary B

Land Biomes

VOCABULARY

After you finish reading the section, try this puzzle! In each of the following items, use the clue to unscramble the letters, and write the term on the line that follows.

1. a region that has few woody shrubs and trees and receives moderate amounts of seasonal rainfall: SARSDANLG

____

2. the layer of soil beneath the surface soil that stays frozen all year round: ROERMFTPSA

____

3. trees that produce seeds in cones: RCOFIEN

____

4. a type of vegetation that includes evergreen shrubs and is located in areas with hot, dry summers: PHARALARC

____

5. plants that keep their leaves all year round: NEVERREGE

____

6. a region that has little or no plant life and long periods without rain: RTEEDS

____

7. a treeless plain found in the Arctic and characterized by very low winter temperatures: RTUAND

____

8. means "to fall off": UUSDDECIO

____

9. found in parts of Africa and home to elephants and giraffes: NVSANAA

____

SECTION SUMMARY

Read the following section summary.

- A biome is characterized by a unique plant community. The plants, in turn, support unique animal communities.
- Plants and animals in a biome are adapted to the climate of the biome.
- Each organism in a biome can be categorized into the ecological role of a producer, a consumer, or a decomposer.
- Deserts are very dry and often very hot. Deserts support plants and animals that use little water.
- Chaparral biomes are fairly dry biomes that support dense patches of shrubs and trees. Animals in the chaparral blend into their surroundings to avoid predators.
- Tundras are cold areas that have permafrost and receive very little rainfall. Tundras support low-growing plants and few animals.
- Grasslands are areas where grasses are the main plants. Prairies have hot summers and cold winters. Savannas have wet and dry seasons.
- Three forest biomes are temperate deciduous forests, coniferous forests, and tropical rain forests.

Name ______________________ Class ______________ Date ____________

Skills Worksheet

Vocabulary and Section Summary B

Marine Ecosystems

VOCABULARY

After you finish reading the section, try this puzzle! Use the clues below to solve the crossword puzzle.

ACROSS

3. animals that form reefs

5. floating raft of algae

6. organisms that float near the surface of marine or fresh water

DOWN

1. rootlike structures

2. where fresh water mixes with salt water from the ocean

4. covers almost three-fourths of Earth's surface

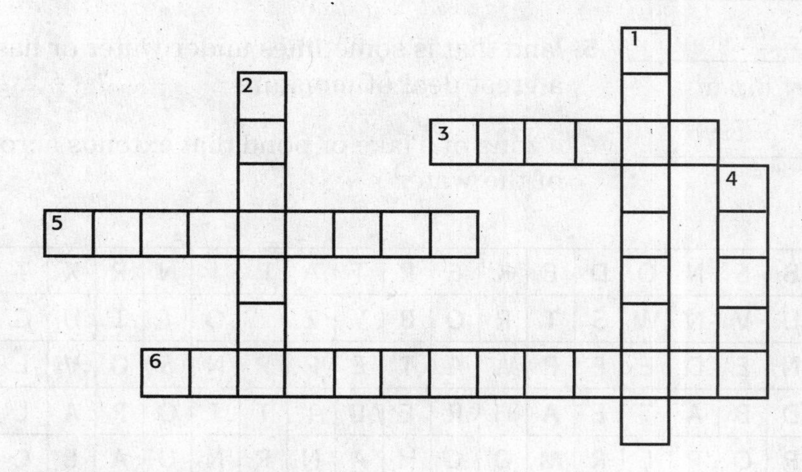

SECTION SUMMARY

Read the following section summary.

- Abiotic factors that affect marine ecosystems are water temperature, water depth, and the amount of light that passes through the water.
- Producers convert solar energy or chemicals from the environment into food, or a form of chemical energy that can be used by organisms.
- Phytoplankton and algae form the base of most of the ocean's food chains. Bacteria that make food from only the chemicals in their environment also form the base of some food chains in the ocean.
- Four ocean zones are the intertidal zone, the neritic zone, the oceanic zone, and the abyssal zone.
- Intertidal ecosystems, coral reefs, estuaries, kelp forests, deep-sea vents, the Sargasso Sea, and polar ice are some marine ecosystems. These ecosystems have unique abiotic factors that support unique communities of organisms with different ecological roles.

Name ______________________ Class ______________ Date ______________

Skills Worksheet

Vocabulary and Section Summary B

Freshwater Ecosystems

VOCABULARY

After you finish reading the section, try this puzzle! In the space provided, write the term described. Then, find those words in the words search puzzle. Terms can be hidden in the puzzle vertically, horizontally, diagonally, or backward.

______________ **1.** the zone of a lake or pond closest to the edge of the land

______________ **2.** a treeless wetland ecosystem

______________ **3.** a wetland ecosystem with trees

______________ **4.** a zone of a lake or pond where no light reaches

______________ **5.** land that is sometimes under water or has soil with a great deal of moisture

______________ **6.** a zone of a lake or pond that extends across the top of the water

B	S	N	O	D	E	K	E	R	T	A	R	I	N	R	X	T
U	V	N	W	S	T	R	O	B	Y	Z	I	O	T	T	U	G
N	E	D	E	E	P	W	A	T	E	R	P	N	S	O	V	L
D	B	A	T	E	A	H	R	E	L	I	T	T	O	R	A	L
R	O	P	L	R	M	O	C	H	A	N	R	N	U	A	L	C
A	N	Y	A	T	Y	K	E	O	N	C	O	L	T	O	I	F
P	Y	A	N	F	A	L	D	P	D	T	E	B	B	R	E	D
H	R	E	D	A	L	B	J	E	K	O	T	O	M	M	E	S
O	A	G	R	P	R	I	N	N	E	L	I	T	O	N	C	A
T	U	S	E	I	G	N	A	W	I	N	M	I	L	O	R	N
O	T	P	I	R	A	L	D	A	F	A	Z	R	E	L	I	N
M	S	I	Z	W	X	R	U	T	A	V	A	E	N	Q	L	A
O	O	W	A	L	E	A	X	E	P	I	G	H	O	U	E	V
L	V	R	E	Y	N	U	E	R	I	S	W	A	M	P	R	A
A	O	L	I	R	O	Q	S	A	R	K	I	P	R	P	I	C
T	L	P	S	A	M	A	R	S	H	A	S	A	N	A	V	E
H	E	N	T	O	E	W	E	H	T	I	H	N	A	B	O	V

SECTION SUMMARY

Read the following section summary.

- Each kind of freshwater ecosystem supports different communities of organisms because each ecosystem has different abiotic factors.
- Organisms can be categorized as producers, consumers, and decomposers in freshwater ecosystems.
- Changes in abiotic factors, such as an increase in sediment and a decrease in oxygen, can cause lake organisms to die. Eventually, further changes can lead to the development of a forest.